NEW   S

Conquering Calculus

The Easy Road to Understanding Mathematics

Other Books by Jefferson Hane Weaver

Conquering Statistics

with Lloyd Motz:

The Concepts of Science: From Newton to Einstein

Conquering Mathematics: From Arithmetic to Calculus

The Story of Astronomy

The Story of Mathematics

The Story of Physics

The Unfolding Universe: A Stellar Journey

Conquering Calculus

The Easy Road to Understanding Mathematics

Jefferson Hane Weaver

PLENUM TRADE • NEW YORK AND LONDON

515
WEA

Library of Congress Cataloging-in-Publication Data

Weaver, Jefferson Hane.
    Conquering calculus : the easy road to understanding mathematics /
Jefferson Hane Weaver.
        p.   cm.
    Includes bibliographical references and index.
    ISBN 0-306-45988-4
    1. Calculus.   I. Title.
QA303.W387   1998
515--dc21
                                                                98-26203
                                                                    CIP

ISBN 0-306-45988-4

© 1998 Jefferson Hane Weaver
Plenum Press is a Division of Plenum Publishing Corporation
233 Spring Street, New York, N.Y. 10013

http://www.plenum.com

10 9 8 7 6 5 4 3 2 1

Printed in the United States of America

This book is dedicated to
Betty J. Ridenour,
with love and affection

Contents

Preface

This book has occupied me for the better part of a year and it is with both reluctance and relief that I bid it a fond farewell. It is a follow-up to my prior book, *Conquering Statistics: Numbers Without the Crunch*, in spirit if not in theme or structure. As with this prior book, I have tried to take a subject which is typically thought of as being sterile and uninteresting and breathe a bit of life and humor into it. Unlike the statistics book, however, *Conquering Calculus: The Easy Road to Understanding Mathematics* is more interdisciplinary and wanders around the mathematical landscape like a child playing along the seashore. At various points, this book considers areas of arithmetic, algebra, geometry, trigonometry, and calculus, cavorting among the basic principles of Euclidean geometry and differential calculus, for example, with equal abandon.

As with statistics, I wanted to be able to write an intelligent but lighthearted book without detracting from the basic integrity of the subject matter. I have not tried to write an authoritative text because I wanted to present the subject in a breezy, indeed whimsical, tone and avoid the tendency of many mathematical writers to lapse into formalistic jargon. Besides, this book is as much a primer as it is a narrative, seeking to outline some of the most fundamental concepts of the major branches of mathematics. Such an approach lends itself to a treatment that is accessible to persons having no real background in mathematics and those who view the subject as something to be avoided at all costs.

The title of this book was intended to suggest that the text meanders through the mathematical countryside, stopping here and there to take in the scenery. Although the calculus is the end result of this journey, it is by no means the only point of interest. There are many wonderful stops in the world of mathematics and mathematical travelers will find a visit to any of these stops to be well worth their time. We may be somewhat presumptuous in using the word "conquering" in the title because one can never really conquer calculus. Perhaps a more appropriate aim would be to help the reader conquer his or her fear of higher mathematics, particularly calculus. Reading this book will not make a novice an expert but it will expose that reader to the general concepts of selected areas of mathematics including the calculus and hopefully instill in the reader a greater understanding of the relevance and importance of the subject.

I have tried to make this book as painless as possible by offering silly examples to illustrate the basic ideas discussed. Many people are intimidated by mathematics because it is a rigorous subject which requires greater study and patience than most other intellectual pursuits. But levity can be a dangerous thing, particuarly when one is trying to discuss the basic ideas of algebra or geometry or calculus. I certainly did not want my tongue-in-cheek examples to be seen as an attack on the integrity of the subject itself nor did I want to create the impression that I viewed mathematics with anything less than great respect. My only hope is that this book will be seen as an enjoyable venture into a subject which has clearly lacked popularity among top-notch publicists for more than two thousand years.

Jefferson H. Weaver

Acknowledgment

This book would not have been possible without the cheerful faces and playful laughs of Mark, Catharine, Matthew, and Caroline Weaver; Haley, Charlotte, and Graham Weaver; Bobby, Allison, and Billy Kahl; and, last but not least, Christy, Jonathan, and Marissa Amill.

Conquering Calculus

The Easy Road to
Understanding Mathematics

CHAPTER 1

Into the Lion's Den

Everything of importance has been said before by somebody who
did not discover it.
—ALFRED NORTH WHITEHEAD

Introduction

Calculus is both one of the most intriguing and one of the most intimidating branches of mathematics to the general public. The very word "calculus" suggests an accessibility that is limited to only a chosen few, typically those who wear glasses with thick lenses and keep their pens and pencils neatly arranged in the plastic sheaths of their breast pockets. Higher mathematics is taught to most high school students beginning with algebra before progressing on to geometry, trigonometry, and, finally, with great fanfare, calculus. So its very position at the end of secondary mathematics education makes it inaccessible to all but the most mathematically inclined students. Because most high schools require only two years of mathematics, most students abandon their mathematical career somewhere between learning Euclid's axioms in geometry and a final, fruitless attempt to stay awake through an entire lecture in trigonometry.

Because many students are extremely busy, what with ethnocentric literacy and mathematics classes, seminars in such vital topics as "self-esteem," days spent competing on school athletic teams,

1

trips off campus for extended lunches, and heated debates as to whether today's students are so oppressed by the modern educational system that they should be treated as political prisoners, there is very little time left to attend actual classes. Admittedly, most high school students are not particularly concerned with the intricacies of higher mathematics because there seems to be a preoccupation with doing only those things that will be "relevant" to their careers or provide an actual "tangible" benefit as they begin their laborious climbs up the ladders of success from the entry-level positions of condiment filler and dining room sweeper through the interim positions of deep-fryer handler and short-order chef to the giddy heights of head cashier and, for those lucky few, a shot at management positions in the second shift. For these bold students it may be difficult to see how a knowledge of calculus will affect the ultimate success of their careers. Quite frankly, their well-meaning mentors would also probably be hard-pressed to come up with a compelling reason as to why they should study calculus.

If one is happy running the milkshake machine or flipping hamburgers or watching restaurant training films in the back room, then, quite frankly, there is little point in acquiring a knowledge of calculus. Oh, one can impress one's fellow shift members by writing various ominous-looking mathematical symbols on the bathroom walls or by tossing such words as "derivative" and "integral" into everyday conversations, but the ordinary demands of a food management career will probably not involve the knowledge of calculus.

But what if you are one of those restless souls who does not see the refilling of the fountain syrup canisters as a sort of culmination of your entire life's work? There are a few people who want something else from life—like a higher education and an opportunity to earn a living wage—and it is these restless souls who should be interested in learning calculus. Sure, there are a number of jobs for which any knowledge above basic arithmetic is essentially a luxury. But there are also many jobs, particularly those in banking and finance, for which a knowledge of mathematics is absolutely necessary. Furthermore, any career dealing with technology or medicine or involving any type of quantitative analysis will require some familiarity with the more advanced branches of mathematics—including calculus.

It would be easy to make a sweeping generalization that everyone must study calculus or risk being left behind in today's technologically advanced industrial society. However, we will always live in a society in which there is plenty of demand for people who will pick up animal carcasses off the highway and staff the guardhouses at the local retirement villages. But it is also true that most of the higher-paying jobs, particularly those in emerging high-technology industries, require some mathematical aptitude. More and more, the ability to converse knowledgeably about quantitative ideas and to understand mathematical concepts is becoming the prerequisite for a successful career.

The importance of the calculus to our society cannot be easily overstated. It is absolutely crucial to the study of chemistry, astrophysics, celestial mechanics, theoretical physics, and economics. Many mathematicians view the calculus to be the most important intellectual tool ever devised by man. It was the calculus which made possible the growth of such technologies as radiation, electronics, atomic and nuclear physics, nuclear energy, electricity and magnetism, space exploration, and much of theoretical astronomy. Indeed, it is difficult to imagine a technology in our society in which the calculus does not play a pivotal role.

Liberal arts majors may quibble with the idea that calculus should be studied purely because more and more of our society's better-paying jobs are in industries in which quantitative analysis is required. But we should point out that William Shakespeare must have agreed with the sentiment expressed in this book regarding the value of studying calculus because he never actually wrote a play in which the protagonist said, "Sod the calculus." Nowhere will you find Hamlet asking himself, "To study the calculus or not to study the calculus—that is the question." Nor will you hear Romeo calling up to Juliet on her balcony, urging her to toss aside her calculus book and run away with him. Indeed, Romeo surely would have asked Juliet to double–check her knapsack to make sure that her calculus textbook was tucked neatly inside along with the dagger and poison that they might require should they decide to end their own lives on a lark.

Of course Shakespeare's failure to complain about the intricacies of the calculus could also be due to the fact that it was not actually invented until some two generations after the Bard's death in 1616. It was in 1665 that the young Isaac Newton retreated to the family farm at Lincolnshire to escape an outbreak of the plague at Cambridge University, where he was attending classes as an undergraduate. While in his temporary self-imposed exile, Newton began to think about the nature of space and time and motion. Over the course of the next 2 years, this young man, who had not been a particularly impressive student at college, constructed a magnificent intellectual edifice, offering the world a vision of a universe governed by immutable physical laws—specifically Newton's three laws of motion (e.g., "For every action, there is an equal and opposite reaction.") and his law of universal gravitation (which was inspired by his being beaned on the head by a falling apple). In doing so, Newton created the field of classical mechanics and posited a universe in which space and time were absolute and the cosmos itself was thought to be akin to a gigantic watch spring wound up by the hand of God.

During this same time at Lincolnshire, Newton also developed the basic principles of the calculus. As with many discoveries in the sciences, Newton's invention of the calculus was prompted by his need to be able to explain concepts such as motion and acceleration in a less cumbersome way than that afforded by geometrical proofs. The development of the calculus proved to be very useful to Newton in his formulation of his three laws of motion and his law of gravity and their application to the study of the motions of the planets around the sun. Newton discovered that the laws of motion and the law of gravity must both be taken into account in order to develop a comprehensive understanding of the workings of the solar system. The orbit of a planet around the sun is "an unchanging geometrical entity that is the same every time the planet revolves around the sun"* whereas "the laws of motion describe the instantaneous change in the motion of the planet in its orbit at any moment."† As the planet hurtles through space around the sun, its state of motion changes continuously because both its direction in its orbital path

*Lloyd Motz and Jefferson H. Weaver, *Conquering Mathematics*. New York: Plenum, 1991, p. 234.
†Ibid.

and its distance from the sun change continuously. (Remember that the planets move in ellipses—not circles—around the sun.) Because the planetary orbits are not perfectly circular, Newton had to grapple with the fact that the force of gravity exerted by the sun on the moving planet would change continuously, which in turn would alter the motion of the planet. Both the direction of the planet's motion and its speed would adjust continuously to accommodate the fluctuations in the force exerted by the sun's gravitational field on the planet.

Newton's genius was to see that if he could specify both the position of the planet relative to the sun and its motion at a given moment, it would be possible for him, using his law of gravity and his laws of motion (which give the rate of change of the planet's motion at each moment), to calculate the position and the motion of that very same planet at a very small interval of time later. He could then repeat these calculations over and over again as the planet continued to move around the sun, until he had obtained a complete picture of the planet's orbital path. Newton thus played the role of a person trying to fit together a gigantic puzzle where each piece revealed the rate of change of the orbit at each moment and the entire puzzle, once assembled, would reveal the complete orbital path of the planet. He realized that he could deduce "the orbit of a planet in this fashion if he could express the instantaneous rates of change in the position and motion of the planet at any moment in terms of the force acting on it at that moment (which also changes)."[*] Newton called these changes "fluxions"; this term was a generic word used to express variations in a given quantity such as motion. He would represent a change in a given quantity by placing a dot over the symbol representing that quantity to show that it would vary over time. If the planet was at some distance r from the sun, for example, then Newton would express the rate of change of r by placing a dot over the $\dot{r}$; if the planet was moving at some velocity v at a given moment, then the rate of change of v would similarly be expressed by $\dot{v}$.

Before we move further into the morass, we should step back and try to appreciate the scope of the task that Newton, perhaps

[*]Ibid, p. 235.

unknowingly, was taking on at the time he began devising the calculus. The German mathematician Felix Klein once said of calculus that "everyone who understand the subject will agree that even the basis on which the scientific explanation of nature rests is intelligible only to those who have learned at least the elements of the differential and integral calculus ..." Klein's sentiments have been echoed by many other scientists and mathematicians because there is no other branch of mathematics that can adequately examine rates of change. Indeed, "it is impossible to appraise and interpret the interdependence of physical quantities in terms of algebra and geometry alone; it is impossible to proceed beyond the simplest observed phenomena merely with the aid of these mathematical tools."* Kasner and Newman, in particular, regard the calculus as both the "cement" and the "implement" for every modern physical theorist who wants to construct a theory about any of the dynamic features of the universe.

Certainly this sentiment would suggest that the calculus occupies a unique position in the hierarchy of mathematics because it *does* deal with change and rates of change. If we were to take a single snapshot of our universe, freezing it forever in time, then we would have no need for a tool such as the calculus because there would be no movement in this universe and, hence, no need for analyzing rates of change. In such a universe we would find that algebra and geometry would suffice for our purposes because they are essentially static. Euclid, for example, who has unfairly been vilified by generations of mathematics students for creating classical geometry, offered a number of axioms that showed how points, lines, and planes could be constructed in an infinite three-dimensional space. But Euclid's work was essentially an unchanging overlay in which these perfect geometrical constructs were superimposed upon our noisy, dirty, imperfect world of everyday experiences. It was frozen in time, much like the snapshot of our picture universe. As such, Euclid's geometry lacked a dynamic quality because it did not take into account the concept of change.

*Edward Kasner and James R. Newman, *Mathematics and the Imagination*. New York: Simon & Schuster, 1940, p. 299.

Euclid's Geometry

The static, immutable nature of Euclid's geometry becomes more apparent if we go back to the third century B.C. where we might have found Euclid (306–283 B.C.) seated on a rock, staff in hand, drawing lines, rectangles, and circles in the sand. Because Euclid had chosen to devote his life to the pursuit of knowledge, he had missed out on the greater opportunities for riches afforded to corrupt public officials. But he did manage to earn a living in Alexandria, Egypt, where the Egyptian kings had built a great center of learning known as the Museum, which boasted what was then the largest library in the world.

Because Euclid has long since left this earth (and cannot maintain an action for defamation of character), mathematical historians have felt free to speculate about his daily life. He is credited with having made the statement "There is no royal road to geometry," which has always struck his admirers as a very profound insight even though it would seem to offer little more than do similar statements such as "There is no royal road to algebra," or "There is no royal road to arithmetic." Even though there is some confusion as to the actual meaning of this statement, many historians have interpreted it as a warning to the intellectually shiftless among us that a true understanding of geometry can come only through hard work. Because Euclid wanted to be remembered for something more than a statement about roadways and mathematics, he compiled a series of thirteen books called the *Elements*, which, next to the Bible, is perhaps one of the most enduring and influential literary works of all time. The Bible can certainly claim to have many more movie adaptations than Euclid's masterwork but the *Elements* has still enjoyed a prolific history in print. "More than a thousand editions have appeared since the invention of printing, and before that time manuscript copies dominated much of the teaching of geometry."[*] Every student who has labored through high school geometry can thank Euclid's *Elements* for having provided the underpinnings of that subject as well

[*]Dirk J. Struik, *A Concise History of Mathematics*, 4th ed. New York: Dover, 1987, p. 49.

as an organizational structure that has deeply influenced the ways in which we pursue knowledge, particularly in the sciences. Sadly, most students fail to appreciate the intellectual sweep and grandeur of Euclid's work or to recognize its achievement in constructing the field of plane geometry from a comparatively small number of basic principles and axioms. This lack of appreciation for Euclid's work is amply demonstrated by the failure of even the most ardent mathematics students to cheer Euclid's name during a lecture or to name their pets after the great Greek.

When Euclid was not drawing figures in the sand or thinking deep thoughts, he amused himself by formulating the basic principles that would eventually become the field of classical geometry. Euclid was not one to allow his thoughts to wander in the clouds; he set about constructing a rigorous body of rules that would provide a logical and self-consistent basis for his work. He began with the most basic concepts of points, lines, and planes and proceeded to build geometry from the ground up. In this way, Euclid used very few arbitrary assumptions (e.g., a point occupies no space, a line extends indefinitely in both directions, a ray is a line with a fixed endpoint extending indefinitely in one direction, only one line can pass through any two points, any three noncollinear points form a plane, etc.) in his work. Although these assumptions lack a certain poetic beauty and do not inspire armies to rush off to war so they can sack undefended cities and kill all the inhabitants, they do provide Euclid's work with both coherence and unity. However, Euclid's geometry did have an air of unreality about it in that it did not seem to manifest readily in the real world. Where could one find, for example, a perfect plane? The floor of a building might be a candidate but closer examination would invariably reveal that it is uneven in places, laden with bumps and valleys that might not be readily apparent to the casual observer. Similarly, one would be hard pressed to find a perfect line. The staffs carried by many of the merchants of that era as they trudged from city to city could be cut in such a way that they would appear to be as straight as a line. But if one looked closely, one would find variations in the smoothness of the surface of the wood due to the patterns of the grain and other imperfections such as knotholes. In short, Euclid offered an abstract world, which, taken by itself, did not appear to be very similar to our own.

Euclid's books are the oldest known complete mathematical texts and are "based on a strictly logical deduction of theorems from a set of definitions, postulates, and axioms."* The first four books of the *Elements* discuss plane geometry and describe the basic properties of lines, angles, and the Pythagorean theorem. The tenth book, perhaps the most difficult and a source of frustration for many generations of mathematics students, discusses quadratic irrationals and their roots. The final three books grapple with solid geometry and feature, among other scintillating topics, the volumes of pyramids and spheres. Euclid's unquestioning acceptance of Plato's philosophical ideas is also revealed in his detailed examination of Plato's theory of forms and his proof that only five geometrical bodies can exist even though the real world is replete with objects with a wide variety of geometrical patterns.

Notwithstanding Euclid's preoccupation with Plato's nonsensical geometric fantasies, Euclid, until the later part of the 19th century, was regarded as the ultimate source of truth in mathematics. His geometry was viewed with awe and, indeed, was more durable than most religions. "Competent and accurate as a measuring tool since Egyptian times, intuitively appealing and full of common sense, sanctified and cherished as one of the richest of intellectual legacies from Greece, the geometry of Euclid stood for more than twenty centuries in lone, resplendent, and irreproachable majesty."†

Moving beyond Euclid's Geometry

To question Euclid's geometry was almost as unthinkable as questioning Jesus Christ about Christianity and, indeed, Euclid was regarded as a sort of divinely inspired rulemaker for classical geometry. But Euclid's magnificent work contained within it the seed of

---

*Ibid., p. 49.
†Kasner, op. cit., p. 134.

future revolutions in mathematics—the fifth postulate—which would eventually give rise to the development of several non-Euclidean geometries in the 19th century. This famous postulate about parallel lines may be stated as follows: "Through any point in the plane, there is one, and only one, line parallel to a given line."* This postulate essentially barred the development of any other type of geometry which would permit parallel lines to merge or to diverge.

The famous fifth postulate was a source of debate among mathematicians, some of whom believed that it was as self-evident as Euclid's other propositions, and others of whom felt that it was not readily deducible from basic geometric principles. Yet this was a debate that did not engage the attention of the lay audience. There is no record that any of the Crusades was launched to defeat the infidels who would dare question Euclid's fifth postulate nor did Charlemagne ever lead his armies into war for the purpose of defeating a free-thinking duke or lord who dared to propose a non-Euclidean geometry. But the fifth postulate continued to be a source of controversy among mathematicians, albeit one that was carried out in correspondences and polite conversations.

The dispute over the primacy to be accorded Euclid's geometry continued to percolate until it finally came to a boil in the 1830s with the presentation of truly revolutionary papers by the Russian mathematician Lobachevsky and the Hungarian theorist Bolyai. Lobachevsky, in particular, had seen that one could replace the fifth postulate and retain all of Euclid's other postulates and thereby create a completely new geometry. Lobachevsky was driven to his conclusion by his belief that the fifth postulate could not be proved and could not be deduced from Euclid's other axioms. Furthermore, he asserted that one could substitute a different parallel postulate and thus create a geometry which would be as true and as logically consistent as the classical geometry of Euclid.

Now you would think that the mathematicians of that time would have leaped to their feet and shouted "Bravo!" but you would be wrong. Many mathematicians are reserved, analytical folk who

---

*Quoted in Kasner, op. cit., pp. 134–5.

seldom slap each other on the back or shout such things as "Good show!" Whether this apparent inability to be affectionate with each other is the result of childhoods spent with parents who were distant and cold or due to the long-term effects of smoking pipes and wearing tweed jackets has never been scientifically investigated. A more plausible explanation might be the inherent conservatism of most mathematicians that is due, in large part, to their belief that they are the guardians of a very long and rich mathematical tradition. In any event, the mathematicians of that era showed very little interest in this new geometry even though it seemed to provide a comparatively fresh way of looking at spatial relationships. No doubt many of Lobachevsky's colleagues were still reveling in the invigorating new insights to be gained from studying Euclid's work, which, after all, had only been around for some 20 centuries. No doubt they intended to take a good look at Lobachevsky's new geometry in due time, which, if the blistering pace with which they had been studying Euclid's geometry was any guide, would have probably required a few hundred years.

But Lobachevsky was undeterred by the tepid reactions of his thickheaded colleagues. Except for the German mathematician Gauss, who has been credited by historians with having anticipated much of Lobachevsky's work, and the Hungarian mathematician Bolyai, who is also credited, along with Lobachevsky, with having invented the same non-Euclidean geometry, few mathematicians of that era seemed willing to accept the idea that a non-Euclidean geometry could exist at all. But if one considered the idea with an open mind and followed Lobachevsky's reasoning, then his argument became very persuasive. All one needed to do was to substitute the fifth postulate with the following statement: "Through any point in the plane, there are two lines parallel to any given line."

Although Lobachevsky would not have considered himself to be a rabble-rouser, his revision of Euclid's geometry had all the impact of a hurricane upon the pristine geometrical world created by Euclid. But this change did not require the complete demolition of Euclid's work. As noted by Kasner and Newman, many of the fundamental principles of Euclid's geometry remained unchanged including the following postulates:

1. If two straight lines intersect, the vertical angles are equal.
2. In an isosceles triangle, the base angles are equal.
3. Through a point, only one perpendicular can be drawn to a straight line.*

But the substitution of the so-called "parallel" postulate did fundamentally alter certain geometrical relationships such as the total number of degrees in the summed angles of a triangle. In Euclidean geometry, the sum of the angles of every triangle equals 180 degrees. But in Lobachevsky's geometry, the sum of the angles of every triangle is less than 180 degrees. Now Lobachevsky was not more efficient in constructing triangles than Euclid in the same way that some people can fit more words onto a single page of paper than others. Lobachevsky's advantage instead came from his recasting of the physical concept of space itself. For Euclid, a triangle could be drawn on a flat plane much like an art student might draw a figure on a sheet of paper. In Lobachevsky's geometry, however, space is no longer flat so that the plane (sheet of paper) on which the figure is to be drawn is similarly distorted. So any attempt by our art student to draw a triangle whose angles total 180 degrees will be unsuccessful because the terrain (the surface of the paper) is not flat but instead negatively curved like the seat of a saddle. As a result, the triangle will not appear to have straight lines as would be the case in Euclidean space but instead three concave lines connected at their endpoints. The angles of this triangle would naturally be less than 180 degrees.

No doubt you are wondering how this information will help you succeed in the world of real-estate finance. Well, there really is no direct connection between real-estate finance and Lobachevsky's geometry but it did provide mathematicians with some very interesting and creative insights into the nature of spatial geometry. In Euclid's geometry, parallel lines would always remain an equal distance apart from each other, never diverging or converging, no matter how far they are extended through space. In Lobachevsky's geometry, however, parallel lines approach each other as they are extended further

*Kasner and Newman, op. cit., pp. 137–8.

outward but they never meet. Needless to say, Lobachevsky's geometry may provide some interesting metaphors for the way in which the participants in high school dances maneuver around the dance floor. But it is doubtful that he was thinking about such profound things when he was scrawling geometrical shapes on his notepads.

Another interesting feature of Lobachevsky's negatively curved space is that as a triangle increases in area, the sum of its angles decreases so that only triangles which have equal areas can have equal angles. In Euclid's geometry, by contrast, one can have equal angles in triangles of any size. If Euclid were to go into a photocopying store with a drawing of a triangle, he could enlarge or reduce the drawing as much as he liked, perhaps making copies as small as the head of a pin or as large as a house, but the angles of the triangles would remain the same. If Lobachevsky were to wander into that very same store, he would find that the sum of the angles of his triangles would increase or decrease accordingly as he shrunk or enlarged his drawing of the triangle. Needless to say, both Euclid and Lobachevsky would also encounter a second problem when they tried to pay for their copies with drachmas or rubles or even olive oil instead of dollars.

The Wonderful World of Riemann

Lobachevsky's non-Euclidean geometry was not the end of the story of the effort to create new geometries by substituting Euclid's parallel postulate. The German mathematician Georg Riemann, emboldened by Lobachevsky's success, took a contrary point of view and instead proposed the following postulate: "Through a point in the plane, no line can be drawn parallel to a given line." Now this statement may strike you as nothing more than a flippant remark by Riemann to hog some of the credit that was being showered on Lobachevsky by simply tinkering with Lobachevsky's revision of Euclid's geometry. But Riemann's proposal was not simply the result of him saying to himself (in German, presumably) that he could monkey around with the fifth postulate on a lark. In fact, Riemann's work would force mathematicians to grapple with the concept of positively curved space which, as we shall see, was to have profound implications for the study of Euclidean geometry and also provide the foundation for Einstein's general theory of relativity.

Riemann, like many mathematicians of his era, lived a very modest life, plagued by ill health and financial difficulties. Then, as now, mathematicians were not in great demand as entertainers and the study of mathematics did not bring with it great wealth or fame. Riemann himself was cursed by poor eyesight. Yet perhaps it was his failing visual acuity which prompted him to think about the conceptual as opposed to the visual aspects of Euclidean geometry. It may have been this approach which prompted him to rethink the fundamental assumptions underlying Euclid's geometry and to posit the existence of what we now know as positively curved space. Positively curved space? Although it sounds like something out of a science fiction movie, Riemann's geometry was every bit as real as that of Euclid, even though it seemed at first blush to have little relevance to the physical universe. But his geometry did have readily apparent applications to the real world and could be easily understood when superimposed upon three-dimensional objects such as spheres.

So Riemann's inspired revision of the fifth postulate was not born of a petulant desire to one-up Lobachevsky or an alcohol-induced haze but instead a revelation that one could have the flat geometry of Euclid, the negatively curved geometry of Lobachevsky, or a distinct positively curved geometry which might as well be known as Riemannian geometry. As with Lobachevsky, however, the reaction of the mathematics community to Riemann's bold idea was akin to their reaction to someone calculating the number of kilometers between Paris and Berlin—apathy and, in some quarters, hostility. Even in the middle of the 19th century, there were plenty of mathematicians who were not particularly anxious to see Euclid's geometry challenged by a few upstart mathematicians. Indeed, Euclid's work was so thoroughly entrenched in the popular imagination, and the common sense applications of his geometry to the physical world so obvious, that many refused to consider seriously these non-Euclidean geometries.

But Riemann was persistent and ignored his critics. He did not waste his time lobbing verbal hand grenades at his detractors in academic publications and questioning their ancestry or sexual proclivities. Riemann was thoroughly convinced that his ideas were as well grounded as those of his Greek idol and, indeed, his geometry could surely be viewed as belonging to the same family tree as that of Euclid's geometry. In revising Euclid's parallel postulate, Riemann proposed that two parallel lines would not continue to remain a constant distance apart as they were extended infinitely in both directions. Unlike Lobachevsky's lines which would converge but never meet, Riemann's lines invariably converged. In short, Riemann rejected Euclid's parallel postulate and, in doing so, made the subtle distinction between space that is infinite and space that is finite but unbounded.

How can space be finite but unbounded? At the risk of wading too deeply into the field of cosmology, we can speak of a finite but unbounded space if we accept the proposition that the gravitational fields of all objects in the universe are mutually attractive so that all objects, however big or small, are drawn to all other objects. Of course this effect is not always discernible as one considers objects as massive as trillion-star galaxies which lie millions, billions, or even tens of billions of light-years apart. But the important point is that the gravitational fields of all objects in the universe exert a pull on all matter, including light. The gravitational pull of the matter scattered throughout the universe will cause anything, even a ray of light, to bend. The greater the density, the more the ray of light will bend. If the density of the matter is high enough, a ray of light moving outward from a given galaxy will bend so much that it will eventually turn around completely and be unable to escape the bounds of the universe even though the universe may have finite limits. In this way, we would be talking about a space that is finite (that space which contains a finite universe) but unbounded because of the absence of any tangible material boundary.

A more terrestrial example is provided by our own earth. A sailing ship can sail the oceans forever and never hit a continent but its voyage is limited to the surface of our planet. In this way, we can speak of the oceans of the earth as being finite (as they are limited in depth and width and volume) but unbounded (as there are no real limits on how many times a ship might continue to circumnavigate the globe).

Now you must admit that this sounds like an interesting cocktail party topic, which would doubtless seem even more profound as the evening wore on and the hand-to-eye coordination worsened. But Riemann's geometry did raise deep and potentially troubling questions as to the nature of space, which did require some thoughtful analysis by mathematicians of that era.

Perhaps the simplest way to approach this problem of visualizing limited but unbounded space is to deal with the quality of being infinite. Some people believe, for example, that the void in which the starry carousel of our universe is cradled may go on forever and ever in all directions. As such, these persons would assert that space is infinite because it has no end. To say that space is finite but unbounded requires something more in the way of understanding. First, it presupposes that we cannot simply move along our merry way in a straight line forever, with our starting point receding ever farther behind us. Instead we could find ourselves moving along the surface of an imaginary sphere and crossing our earlier paths over and over again for all eternity. Kasner and Newman point out that the paths followed by the hands of a clock also fall within this definition of finite yet unbounded: "Moving in any given direction, like the hands of a clock, we can keep going forever, forever retracing our steps."* The hands of the clock can turn around for all eternity and the paths that they trace will certainly be finite in spatial extent but will be unbounded because of the absence of any physical barriers. Unlike Euclid and Lobachevsky, however, Riemann offered a geometry in which any number of perpendicular lines can be drawn from a point to a given straight line.

But how can this be true? Perpendicular lines intersect straight lines at perfect 90-degree angles. The answer may be found if we look at the lines of longitude and latitude on a globe. The longitudinal lines begin at a single point on the top of the globe (the north pole) and then gradually spread apart as they move southward until reaching the equator where the longitudinal lines spread out to their maximum extent. As the lines extend further southward below the equator, they begin to converge once again until reaching a single point at the southern pole as shown in Figure 1. A globe divided

---

*Kasner and Newman, op. cit., p. 139.

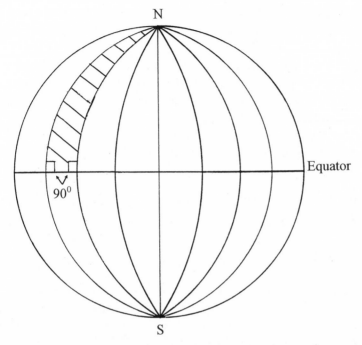

N

Equator

90⁰

S

Figure 1. In positively curved space, the sum of the angles of a triangle exceeds 180 degrees, as shown by the shaded triangle.

along its longitudinal lines resembles an orange that has been sliced along its north–south axis. (The beauty of this model is that it is not limited to oranges but can be extended to all varieties of citrus fruits.) If we examine this model from a geometrical standpoint, we start to see that Riemann was not quite as crazy as some of his colleagues alleged.

Because the longitudinal lines all converge at polar points, they must all necessarily approach each other as they move farther away from the equator, regardless of whether they are extended to the north or the south pole. But we start to see that a curious thing occurs, which was not anticipated by Euclid's plane geometry. We recall that Euclid solemnly declared that the sum of the angles of any triangle was equal to 180 degrees. It did not matter one bit to Euclid if the triangle was the size of an elephant's toenail or the Sahara Desert; the angles would still total 180 degrees. But Euclid did not offer a geometry in which the triangles were overlaid on a curved surface as is the case with the longitudinal lines on our globe. Because Euclid was concerned with purely abstract figures, whether triangles, circles, or rectangles, the planes in which he described these figures were similarly abstract (flat). Euclid thus did not consider surfaces of varying topological features nor did he consider surfaces having negative (Lobachevsky) or positive (Riemann) curvature. His planes were as flat and unvarying as the proverbial glass pond.

Riemann, like Lobachevsky, realized that the revamping of the parallel postulate would force one to consider non-Euclidean spaces. Unlike Lobachevsky, whose triangles could never boast a total of 180 degrees, Riemann found that the sum of the angles of a triangle in positively curved space could never be as little as 180 degrees. The reason for this became apparent if one returned to the trusty longitudinal lines of a globe and began studying the angles of any of the triangles formed by extending two lines from a common polar point to the equator. The vertex of the triangle at the pole has a certain number of degrees. But we find that the two longitudinal lines intersect the equator at right angles so that the interior angles of this triangle at the equator are each 90 degrees. Those readers who have not yet tried to return this book will recognize this apparent violation of Euclid's declaration that the angles of every triangle total 180 degrees. After all, we have two 90-degree angles at the equator and then an additional angle consisting of an unspecified number of degrees at the pole. As a result, the total number of degrees in the angles of a Riemannian triangle is greater than 180 degrees. So Riemannian geometry permits any number of perpendicular lines to be drawn from a single point (the pole of a globe) to a given straight line (the equator). We can see the truth in this statement if we retrieve a sharp knife and an orange and begin slicing an orange along its longitudinal and equatorial lines.

The cut along the equatorial line of the orange necessarily creates two halves, either of which can be used for our purposes. We then begin cutting slices from either half, beginning at the polar point of the orange and continuing down to the equator of the orange. As we cut more and more slices from a given half, we find that Riemann's suspicion about being able to pass any number of parallel lines through a given point (the pole) appears to be more than the mad rants of a German mathematician because each cut of the knife allows us to create an additional parallel line which passes from the pole to the equator. If we look at each of the slices, we will find that they appear to be similar in shape to the triangles formed on the surface of the globe. The edges of the slice along the equator together appear to be equal to 90 degrees each or a total of 180 degrees. The other end of the slice has a much smaller interior angle but it still measures some number of degrees. Even if we take a given slice and slice it lengthways one, two, three, four, or even five more times, the interior angles of the triangle formed by our frantic carving activities will still be greater than 180 degrees.

Yet there is something troubling about this analogy, even for those of us who consume citrus products with great gusto. We have trouble seeing how the lines that form these slices can be said to be straight. If we go back to our globe and we run our single remaining unbandaged finger along two intersecting longitudinal lines, we have to wonder whether it makes sense to speak of these lines as actually being straight. After all, these lines are superimposed on a curved surface and, as a result, have a degree of curvature commensurate with that of the surface.

Any mathematician you find hanging out on a street corner will be happy to point out that the adage about the shortest distance between two points being a straight line does not necessarily apply in geometry. In fact, the definition to be accorded to the word "straight" will vary, depending on the particular surface being considered. If we are on a flat plane of the sort offered by Euclidean geometry, then a straight line will be formed by connecting any point to any other point. In Euclidean geometry, the shortest distance between two points is a straight line. We can certainly see the truth in this statement if we imagine ourselves to be standing at one point, which coincidentally happens to be the tavern where we have spent the last several hours imbibing numerous pints of beer and engaging in deep philosophical conversations with fellow social drinkers whose statements seem more and more relevant to the human condition as our own blood alcohol levels increase. Now that the bar has closed, however, we must travel to the hotel where we are staying, where we know that there is a refrigerator laden with candies, crackers, and enough liquor to wash down our midnight snack. Because we are all greatly concerned with being efficient in our actions and not wasting a lot of time, we do not want to take a meandering path back to our hotel. We want to take the most direct route in the shortest possible time so that we do not sober up and have to confront reality before we get to our rooms. If our hotel is seven blocks south and four blocks east of our current position, then the most direct route would involve a diagonal path which would necessitate walking through the walls of numerous structures. Because we cannot pass through masonry without suffering extensive injuries, the diagonal route is not preferable from a realistic standpoint. Even though Euclid would have probably urged us to continue hurtling our battered, bloodied bodies at the wall of the first building blocking our diagonal path, we would ultimately be forced to tell Euclid to mind his own business and to take the shortest *available* straight line path between the tavern and our hotel. As a result, we would walk (or stumble) seven blocks south and four blocks east until we reached our hotel. This journey would be akin to tracing a line between two points in Euclidean space because we would make our way along a flat (or relatively flat) surface until we reached our destination.

For those readers wondering about the relevance of this discussion, we should point out that the concept of "straight" is contextual. When we consider a positively curved surface such as the surface of a globe, a straight line is one which proceeds along the surface of the globe from one point to another. A longitudinal line, for example, is a straight line on the surface of the globe and would be considered "straight" in the positively curved space offered by Riemann. But Euclid would have likely thrown a tantrum if we tried to insist that he should consider such a line as being straight if drawn in Euclidean space because such an argument, quite clearly, would be wrong.

But we should not confuse the issue of "straightness" with that of "shortness." The concept of straight is dependent on the topology of the surface along which the line is drawn. However, the shortest line is clearly a straight line drawn between two points. On a globe, we can draw a straight line along the surface from the pole to the equator. But it is not the shortest possible line because we could begin at the pole and tunnel head-on in an unchanging direction through the mantle of the globe until surfacing at a point on the equator. This tunnel would have a shorter linear length than the distance one would have to travel along the surface of the globe to move from the pole to the equator. But only the line drawn along the surface of the globe would be considered "straight" in a topological sense. Needless to say, few people would want to go to the trouble to bore a tunnel through the earth in order to have the satisfaction of knowing that they took the straightest possible path between the north pole and a point on the equator. After all, it would clearly be much simpler to climb aboard a dogsled and travel across the surface of the earth even though its curvature would preclude our hardy travelers from being able to claim that they took the shortest possible route.

Regardless of whether we are playing with the geometries of Euclid, Lobachevsky, or Riemann, we will often find it helpful when dealing with mathematical problems to determine the shortest distance between two points. As Riemann's geometry conceives of space as positively curved and Lobachevsky's geometry features negatively curved space, it does not require an incredible intellectual effort to view the shortest distances between two points on these surfaces as curves. In Euclidean geometry, of course, the plane is flat, so a straight line extended through that plane is flat and has no curvature. With non-Euclidean geometries, a straight line extended through a plane cannot connect two points lying within that plane. This point will become more apparent if we reach into our bag of analogies and offer the following example: Suppose that we have a cylindrical lampshade which we just purchased at a garage sale. We were drawn to this lampshade because we are collectors of lampshades and this particular one was in very good condition, boasting a perfect roundness and a spectacular brown, green and orange painting of a rat. Once we get home with the lampshade, however, we find, upon further examination, that the layers of the shade are actually separating. Not wishing to see our prized possession become more frayed, we retrieve the biggest needle and spool of thread we can find. As we begin sewing the layers of the shade back together, however, we discover that our needle can draw a thread between two points but that the thread, once pulled taut, does not remain flush against the surface of the lampshade but instead lies a tiny bit above the surface of the lampshade. The thread thus creates a straight line which connects the two points but which, due to the curvature of the lampshade, does not traverse the surface of the lampshade.

This point may be more clearly illustrated if we imagine that Sir Reginald Phipps and his hardy band of explorers have just finished building a rope bridge so that Sir Reginald can cross a deep but fairly narrow river which happens to be infested with crocodiles. As some of the lower-ranking members of Sir Reginald's party have already managed to get to the other side by crossing a narrower part downstream and losing only a few limbs in the process, they now wait patiently for Sir Reginald's men to load one end of the rope bridge into a makeshift catapult. Once the catapult is launched, the rope bridge shoots across the river and is grabbed by the people on the other side. Initially, the rope bridge is lying in the water and is not taut. The rope bridge is thus lying on the surface of the earth but it is not a straight line. Once the bridge is pulled up and straightens itself out, then it will no longer be lying in the river but instead will extend from one riverbank to the other and will be above the river itself. So the bridge in its taut form will represent the shortest line between the two points from where it begins to where it terminates. It will not be "straight" in a topological sense because it does not lie in a flat plane but instead lies within a plane that has some degree of curvature—in this case, positive and negative.

Euclid was obviously not a sailor because his geometry assumed that the surface of the earth is flat. Had he ventured out to sea and watched the shore recede beyond the horizon he might have reconsidered his ideas about the perfect flatness of geometrical planes. But the sailors of Euclid's day undoubtedly had some suspicion that the earth was not perfectly flat, as was commonly assumed, because of the limitations on their ability to see across the ocean: "[A]n ocean is not thought of as a flat surface (Euclidean plane) if even moderate distances are concerned; it is taken for what it very approximately is, namely a part of the surface of a sphere, and the geometry of great circle sailing is not Euclid's. Thus Euclid's is not the only geometry of human utility.... So, in the highly practical geometry of great circle sailing, which is closer to real human experience than the idealized diagrams of elementary geometry ever get, it is not Euclid's postulate which is true—or its equivalent in the hypothesis of the right angle—but the geometry which follows from the hypothesis of the obtuse angle."* In other words, Riemann would have probably been a much better sailor than Euclid even though his pale skin would have suffered severe burns on voyages to the tropics. More to the point, Riemann's geometry of positively curved space is, at least for sailors, the more relevant geometry, because it takes into account the fundamental curvature of the earth's surface.

But Euclid's geometry is of great importance to humanity. Even though there is no such thing as a perfectly flat surface on the earth or a line that can be extended to infinity in both directions, Euclidean geometry provides us with a conceptual map to help us organize our perceptions of the world. Needless to say, the ability to geometrize was of critical importance in the construction of roads, buildings, ships, and nearly every other type of technological improvement. It was much easier to build a house using bricks cut in rectangular shapes than, for example, bricks shaped like stars because even the dimmest of do-it-yourself homebuilders could see that a wall of rectangular bricks would have few gaps or holes. A house built of star-shaped bricks would be permeated with many openings and crevices and would not provide watertight shelter. Furthermore, the rectangular bricks would fit together more tightly and thus provide a stronger buffer against the elements.

One can use Euclid's geometry at a very basic level to solve everyday problems. We shape bricks in the shape of rectangles because we know from Euclid's descriptions that they will fit more tightly together than other, more irregular shapes that have fewer surfaces which can be placed in contact with other, similarly shaped bricks. Euclid's description thus provides us with a concept of the rectangle which we utilize in the making of bricks. We know that we are incapable of making a brick that is a perfect rectangle because we cannot, no matter how carefully we scrape and buff the sides of a

*Eric Temple Bell, *Men of Mathematics*. New York: Simon & Schuster, 1937, pp. 302–3.

given brick, fashion one that consists of perfectly flat surfaces that intersect at precise perpendicular angles. Even the pickiest brick maker in the world will not be able to make a brick that will rival the perfection of Euclid's rectangle because we simply cannot approximate the supreme elegance of Euclid's concepts. But it is Euclid's unattainable ideal that provides the brick maker with the intellectual direction he needs to shape bricks that can form the strongest wall using the least amount of material.

It is this intuitive sense that we all have about Euclidean geometry that encourages us to try to approximate the ideal shapes offered by the *Elements* in all of our daily activities. In walking from one place to another, for example, we will usually try to take the shortest possible route, which will involve one or more different straight-line pathways. If we are hurrying over to a zoo to throw rocks at the most recent shipment of endangered animals, for example, we will not want to take a wildly meandering route which will take us far out of our way and add much time to our journey. No, those of us who view the beaning of whooping cranes and bald eagles as something akin to a religious experience do not want to risk losing even an extra few minutes simply to take a circuitous route so that we might take in some of the sights and sounds that the warehouse district of any crumbling city might offer. As a result, we would want to take the shortest possible route so that we might resume our continuing quest to understand fully the relationship between mediocre hand-to-eye coordination and senseless carnage.

Some historians of mathematics believe with some justification that Euclid has been given too much credit for the contributions he made to mathematics whereas Lobachevsky has been slighted by posterity. The American mathematician Eric Temple Bell, for example, has offered a view of Euclid's contribution to the *Elements* which some would say is less than charitable: "His [Euclid's] part in the *Elements* appears to have been principally that of a coordinator and logical arranger of the scattered results of his predecessors and contemporaries, and his aim was to give a connected, reasoned account of elementary geometry such that every statement in the whole long book could be referred back to the postulates. Euclid did not attain this ideal or anything even distantly approaching it, although it was assumed for centuries that he had."* Bell would thus agree that Euclid's role was more that of a synthesizer than of an originator. But Euclid's place in the pantheon of mathematical genius is still secure even if the originality of much of his work can not be proven because he organized all of the principles of geometry into a coherent compendium.

A Dash of Lobachevsky and Copernicus

Euclid's, Riemann's, and Lobachevsky's geometries, do share a common characteristic in that they are all eminently practical in the world of our daily experiences. "For any everyday purpose (measurement of distances, etc.), the differences between the geometries of Euclid and Lobachevsky are too small to count, but this is not the point of importance: Each is self-consistent and each is adequate for human experience. Lobachevsky abolished the necessary 'truth' of Euclidean geometry. His geometry was but the first of several constructed by his successors [such as Riemann].... For some purposes Euclid's geometry is best or at least sufficient, for others it is inadequate and a non-Euclidean geometry is demanded."†

Like Riemann, Lobachevsky's greatest contribution to mathematics may have been his stubborn refusal to accept Euclid's postulates as being immutable. He could claim a healthy skepticism for the prevailing opinion among mathematicians of that era that Euclid's work represented both the beginning and the end of geometry and that the *Elements* was "an absolute truth or a necessary mode of human perception in his system of geometry."‡ As such, Lobachevsky dared to challenge the validity of a paradigm that had dominated mathematics and mathematical thought, exhibiting a degree of daring and courage that prompted some to compare him with Copernicus. However, that sort of comparison may do less justice to Lobachevsky than to Copernicus because Copernicus essentially dusted off a heliocentric model of the universe which had been

*Ibid., p. 299.
†Ibid., p. 306.
‡Ibid.

proposed by the Greek astronomer Aristarchus and offered it as an alternative to the clumsy, epicycle-laden geocentric cosmology which had been proposed by the Egyptian Claudius Ptolemy. Certainly Copernicus was a bold thinker in daring to challenge the cosmological orthodoxy which was embraced by the Church, particularly when it was not uncommon for such imaginative persons to be branded "heretics" and burned at the stake. But Copernicus was not an original thinker; his genius was to repackage Aristarchus's work in a more updated form. Moreover, Aristarchus had assumed that the planets move in perfect circles around the sun and Copernicus did nothing to challenge this assumption. The Copernican system also featured circular orbits, a flaw which was remedied only after the mathematical astronomer Johannes Kepler spent much of his life carrying out tedious calculations which showed that elliptical—not circular—orbits were the only correct models that agreed with the existing base of astronomical observations.

Lobachevsky also challenged the prevailing orthodoxy in mathematics but revamped Euclid's parallel postulate, paving the way for future revolutionaries such as Riemann. But Lobachevsky did not need to destroy Euclid's geometry in order to devise a new geometry. By comparison, Copernicus had no choice but to offer a completely different cosmology from that of Ptolemy because one had to choose between either an earth-centered or a sun-centered solar system. One could not tweak the orbit of Mars, for example, and somehow reconcile the two models. One had to either accept Ptolemy's model as an article of faith or kick in the proverbial door as did Copernicus.

But even though we are inclined to cheer Lobachevsky's daring proposal that space may be negatively curved like the saddle of a horse and that the sum of the angles of a triangle can never equal, let alone exceed, 180 degrees, his geometry is a little more difficult to imagine than that of Riemann or Euclid. Instead of the robust triangles shown in Figure 1 which reek of vim and vigor, a triangle in negatively curved space will strike one as being anemic and run-down. Figure 2 illustrates the differences in shape among three equilateral triangles in Euclidean, Riemannian, and Lobachevskian space. The first triangle would be found in Euclidean space and the sum of its angles would be equal to 180 degrees; the second, more

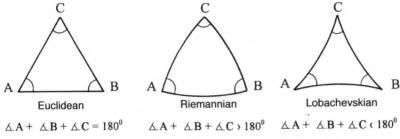

Figure 2. Triangles in Euclidean, Riemannian, and Lobachevskian space.

muscular triangle would hail from Riemannian space; the third, sickly triangle would call Lobachevskian space home. Although these triangles certainly differ in their properties, their differences are due solely to the topology of the surface on which they are placed.

Now you may be ready to consign Lobachevsky's geometry to the scrap heap because of the difficulty of imagining a negatively curved space. But this would be a mistake because Lobachevsky's geometry, like that of Riemann's geometry, provides a possible model for the structure of the universe. In a space consisting of negative curvature, all the objects of the universe are moving apart from each other. Unlike the positively curved Riemannian universe, however, the objects in a negatively curved Lobachevskian universe will not converge upon each other but will continue receding from each other forever because there is not a sufficiently strong mutual gravitational attraction among the stars and galaxies to brake this expansion. So the Lobachevskian universe would continue dissipating itself for all time and all the stars would eventually burn out and become as cold and dark and lifeless as the vast surrounding void.

Neither Riemann nor Lobachevsky knew anything about cosmology but it is eerie how relevant their mathematical models have been for the study of the structure of the cosmos itself. This is but one example of the way in which the world may be said to mimic mathematics (and mathematics the world) because it is the concepts of mathematics that make it possible for us to organize our thoughts about the physical world and formulate theories from which useful conclusions can be drawn.

Despite the theoretical value of these alternative geometries, the fact remains that our intuitive sense about the nature of the world tends to embrace Euclid's vision in which all surfaces are perfectly flat. There is no compelling reason why this should be the way we view things but we cannot ignore our kneejerk tendency to think of flat areas of land and water as being perfectly flat. We stand on the shore of the sea and look toward the horizon and we perceive the ocean before us as having no curvature. But we know that the earth is curved and that the horizon is where it is because of the earth's curvature. Yet we still think of the surface of the ocean as being as smooth as a gigantic pane of glass. For any inhabitant of a planet, a true geometry will clearly mimic that offered by Riemann more than the geometry of Euclid. But our everyday perceptions do not allow for the more complex positive curvatures of surfaces offered by Riemann because the amount of curvature is not perceptible when we are dealing with the distances from our homes to our offices. Even if we climb aboard an airplane and stare downward at the earth five miles below us, we will not fully appreciate the curved surface of our planet (perhaps because we might have our faces buried in airsickness bags). Needless to say, the distances that we walk—whether around the neighborhood or along a highway thumbing for a ride from a passing psychopathic killer—are not so great that we are going to be able to appreciate the curvature of the earth's surface.

So even though we know that the earth is round and that it is possible for any number of lines to emanate from a polar point to perpendicular intersections with the equator (score one for Riemann), Euclid (perhaps because of our affinity for celebrities with one name) will always be the sentimental favorite because his geometry is the least upsetting to us. When we sit down at a desk with a pencil and ruler in hand to draw a straight line, we derive some measure of comfort knowing that we are drawing what appears to us to be a straight line. Imagine the consternation that would result if we knew that we would have to adjust even the shortest straight line by some amount to account for the surface curvature of the earth to please Riemann. What would be the point of even bothering trying to draw a straight line if you knew that it was doomed to be inaccurate from the start? Fortunately, we are able to ignore the curvature issue and draw Euclidean lines on the surfaces of Euclidean desks with Euclidean pencils. This is not to say that these straight lines are completely correct, since we do live on the curved surface of a spherical planet, but Euclid's geometry will suffice for those of us who do not wish to try to draw slightly curved lines with slightly curved rulers. After all, it is easier to use and to understand than either of the geometries offered by Lobachevsky or Riemann.

Perhaps the most important point to be gleaned from this chapter is that all branches of mathematics provide abstract visions of the real world. Mathematics provides a blueprint whereby we can quantify the apparent chaos of our own world. In the case of geometry, the blueprint helps us to understand spaces and forms. But geometry has its limitations because it cannot express dynamic actions such as velocity or acceleration or motion in any meaningful way. It was this need for a "dynamic" mathematics which led Newton and Leibniz to develop the calculus. But this development did not occur out of thin air; it represented the culmination of hundreds—even thousands— of years of intellectual endeavor in which reasoning processes based on rigorous logic were developed and various branches of mathematics were devised to deal with different types of real-world problems. Before we can deal with the mathematics of the calculus and some of the problems it was designed to solve, however, we will first need to consider both the thought process which made possible its development as well as some of the other branches of mathematics which provided important stepping stones in the development of the calculus.

CHAPTER 2

Of Mathematical Reasoning and the Mathematical Mind

Here and elsewhere we shall not obtain the best insights into things until we actually see them growing from the beginning.
—ARISTOTLE

Introduction

To meander through the wilds of mathematics is to engage in a journey for which one must make certain basic preparations. Unlike a safari through the African subcontinent, however, one does not need to take 50 boxes of supplies or hire an entire village of stewards to help carry the load. No, it is possible to travel through the densest thickets of mathematics using only a pencil and paper and an active imagination. But one cannot merely plow ahead and expect to be able to master the intricacies of algebra, geometry, or even the calculus without having some awareness of the intellectual process by which mathematical concepts are formulated and tested. Because mathematics is a rigorous discipline and perhaps the most conceptually based of all the sciences, one must understand the reasoning process that guides the development of all mathematics. Not coincidentally, it is the same type of logically based reasoning process that is used by philosophers to grapple with the "ultimate questions" of life and existence.

31

The world is full of persons who are content to float along through life and not strain their intellects too much by thinking about anything more than the evening schedule of television programs. Fortunately for the future of civilization there are some intellectual adventurers who are not content simply to drift along waiting for their steadily rising cholesterol levels to one day kill them. These people (who, by coincidence, have purchased most, if not all, of the author's books) give serious consideration to some of those "ultimate" questions which have bedeviled philosophers from the beginning of recorded history.

When we refer to "ultimate" questions, we are talking about those questions that arguably lie outside the purview of science and mathematics such as: "Why are we here?" "What lies beyond the universe?" and "What is the meaning of life?" When considering such topics, we do not address them expecting to come up with a pithy answer because these questions are not the type that typically lend themselves to a concise resolution. Instead they are questions that are intended to provoke one's thoughts and jumpstart the intellectual reasoning process. Even though these questions may lead us to consider issues relating to the practical concerns of everyday life, they are more valuable as a means to give flight to our creativity. But the usefulness of such questions lies in the logical consistency of the reasoning process used to consider these issues. Here is where these ultimate questions share common ground with those questions which are geared toward eminently practical issues such as designing a better mousetrap or, if you are a geneticist, a better mouse (such as one who is lactose intolerant so that he will leave the cheese in the cupboard alone), or even a more useful branch of mathematics such as the calculus. Addressing the ultimate questions of philosophy or the pragmatic questions of everyday life both require a clear-headed, consistent approach and, indeed, will only yield anything of value if the investigator attacks the problem by using a rational form of inquiry.

Types of Reasoning

Those of us who tower above mere mortals in intellectual capacity know full well that there are three types of reasoning by which a problem may be considered—reasoning by analogy, inductive reasoning, and deductive reasoning. Of the three, reasoning by analogy is the least satisfactory from a logical point of view because it involves comparing the characteristics or features of one situation with that of another and using the results of the first situation to draw conclusions about the likely outcome of the second situation.

Suppose that I watch the famed demolitionist Clyde Hackett wire a condemned building with dynamite and blow it up with a single push of the plunger so that it falls in on itself, kicking up a massive cloud of dust. Knowing that Clyde is something of an imbecile when it comes to most matters, I might decide that as I did graduate from trade school and have spent my entire career working as a bookkeeper for a local bank I am eminently qualified to begin a career in demolition. My strong feelings about my prospects for success might be bolstered by the fact that I always enjoyed knocking down the block cities of other children when I was in kindergarten and setting fire to paper on a sidewalk with a magnifying glass. So my walking into an explosives shop by the local schoolhouse and purchasing a few boxes of dynamite would be prompted by my reasoning by analogy. In following such an approach, I am not attacking a specific problem in an intellectual manner but am instead considering the circumstances surrounding Clyde's success in the demolition business and then concluding that I, too, can also successfully blow up buildings based on my admittedly subjective comparison of my abilities and accomplishments with those of Clyde. When reasoning by analogy, there is obviously nothing in the way of a precise intellectual reasoning process involved because it is more akin to trying to stuff a round peg into a square hole or, in our example, trying to define the central elements which figure into Clyde's success (other than extremely lax munitions laws) and then trying to find those very same elements (or at least most of them) in my own situation. Of course my concluding that I am eminently suited for placing explosive charges in public buildings necessitates that I consider that any obvious differences between our respective situations are not relevant. The fact that Clyde likes fried peanut butter and banana sandwiches (which I cannot stomach) probably has little relevance to whether I can be a successful demolitionist. But I might be more concerned if I found that Clyde had received extensive training in explosives while working as a counselor at the local youth camp whereas I might have never received any such training aside from an occasional self-help mailing from a local survivalist group. My lack of formal training in building explosive devices might be a significant enough difference to make me wonder whether in fact I should abandon my bookkeeping career. So my reasoning by analogy might be a good approach or it might be the reason that I would one day be known affectionately among the local school-children as "Three Fingers." Of course I could always ignore this difference in our educational backgrounds but such a course would probably be at my own peril.

There are certainly many other examples of the pitfalls associated with reasoning by analogy. Suppose that my good friend Oscar Klein, who had practiced ballet since he was three years old, had recently been admitted to the exclusive Fluge Academy of Clog Dancing in New York City. Oscar had been a devoted student of all forms of dance and had spent six hours a day clogging under the watchful eyes of the world's best known cloggists, gearing his entire life toward the one day when he would receive that very special letter from the Fluge admissions office inviting him to study under Generalissimo Erwin Fluge (*ret'd, Luftwaffe*) for a grueling four years in the hopes of joining the professional clogging circuit and touring the world for the rest of his career. Now I might watch Oscar banging his shoes on the dance floor during a clogging rendition of *Swan Lake* and think that there really is nothing to this clogging routine. I might then purchase a pair of clogs from the local lumberyard and begin practicing all of the classical routines such as the *Nutcracker* and *Lassie*. After a solid two weeks of practice with daily clogging routines ranging up to twenty minutes, a duration befitting one who is serious about studying the arts, I might believe that I have sharpened my clogging skills and perfected my dance routines so that I am now ready to be considered for admission to Fluge. Of course most people who can claim at least a tenuous hold on reality would wonder how I could even hope to be able to audition for Generalissimo Fluge when I have no formal training or clogging experience. But because I am so deeply in love with myself and feel I can do no wrong, I ignore the differences between the dance training received by myself and Oscar and go visit the Fluge Academy.

Not surprisingly, things do not go quite as well as expected. Although I am able to secure a five-minute audition with Fluge himself, tragedy strikes when, while doing the can-can clog, one particularly vigorous kick sends my left clog hurtling through the air with the speed of the stone that felled Goliath. Fluge then makes the mistake of standing up to order me off the stage and gets smacked in the face by my shoe. He then drops to the ground like a sack of cement and his toady assistants run to the telephone to call the paramedics. After being hustled out by two security guards (*sans* left clog) and tossed headfirst onto the sidewalk in front of the Academy's doors, I then have a quiet moment before passing out to contemplate whether my reasoning by analogy in this situation may have been flawed. Was my situation sufficiently similarly to that of Oscar to justify my conclusion that I could dance at the Fluge Academy or had I, in my giddiness over dancing at the most prestigious clogging institute in the world, conveniently ignored some possibly important considerations that would have undermined my attempt to analogize between Oscar's situation and my own? No, my complete lack of formal training had nothing to do with my being tossed out of the Fluge Academy. Oscar had obviously decided that he could not stand to have me at the same Academy clogging away. Undoubtedly, such a prospect threatened both his fragile ego and his prospects for a professional clogging career so he had to have convinced Generalissimo Fluge to toss me out like a bucket of bathwater. So it was a conspiracy that reached into the highest levels of the clogging pantheon which denied me the opportunity to pursue a clogging career—not my alleged shortcomings as a dancer.

Reasoning by analogy is fine for making snap decisions because it is clearly not a particularly rigorous intellectual process. As with my decision to pursue a career in the demolition business, however, the reasoning process depends totally on the degree to which the circumstances have identical characteristics or features as well as the reasoner's ability to focus on the differences between the two situations. Because of these problems, we need to consider another type of approach that can be used to solve problems—reasoning by induction.

Reasoning by induction differs greatly from reasoning by analogy because the inductive process involves making repeated observations of some event and then at some point concluding that that very same event will always occur under those very same conditions. Suppose that I lived next door to the "Leaping Lobombos," a high-wire circus troupe that likes to practice aerial routines by leaping off the roof of their house and doing twists and spins and twirls through the air before landing headfirst on a bed of nails. After watching the various members of the ensemble calmly leap into the air and impale themselves on galvanized nails, I might conclude that every time one of the Leaping Lobombos takes the plunge, the trauma unit at the local hospital will be receiving a new patient. I might find after five or six such occurrences that the leap off the roof was invariably followed by the visit by the ambulance. Fancying myself as having a rigorous scientific mind, I would, through inductive reasoning, conclude that a leap by a Leaping Lobombo off the roof of the Lobombos' home would always be followed by a visit from an emergency rescue vehicle. So it is the fact that I repeatedly observe the same outcome that causes me to conclude that the same outcome will always occur under those circumstances. This leap from repeated observations to a conclusion that holds true at all times is the basis for inductive reasoning.

Although inductive reasoning would appear to make a great deal of logical sense, it does have its limitations. Quite simply, there is no guarantee that the same result will occur every single time. To return to the Leaping Lobombos, it may be that every so often one of the family members will miss the bed of nails on landing and not need extensive medical treatment. As a result, our hypothesis that a leap off the roof of the Lobombos' house will invariably be followed by an ambulance trip to the hospital may not be true; out of every 100 leaps, there is still that one nagging exception to the rule that prevents us from saying that the ambulance will always start its engine and turn on its siren when it gets an injury report from the Lobombos' household. But inductive reasoning does clearly have an advantage in that it enables us to conclude, upon making a given number of observations of a particular phenomenon, that the outcome of those experiments will always occur. This conclusion may be wrong sometimes but it does provide us with a sort of mental shortcut that enables us to get on with our daily affairs.

How can inductive reasoning be helpful to us? Let us assume that we are innocent in the ways of the world and have no experience handling razor-sharp cutlery. If we see our Cousin Joe accidentally cut his finger while trying to carve the Thanksgiving turkey and we watch the kid down the street scrape off part of his forearm while trying to cut some intricate parts for his scaled airplane model and we hear that the handyman who is replacing some drywall on the back of our house severed his foot with a chain saw, then we may conclude, with some justification, that sharp objects can cause harm to us. We can test this theory further (perhaps during the halftime ceremonies of a professional football game when we are sharing a single brain with our drinking buddies on the living room couch) by using the point of a sharp knife to see who has the toughest skin. It may be that even though most of us will be bleeding profusely all over the house, there will be a single exception to the general belief that a plunging knife will result in a gaping flesh wound. If Uncle Stanley, for example, who has extremely calloused hands from his many years as a farmer, plunged a knife on a dare into his hand, then he might not suffer any injury at all. We would all be amazed that Uncle Stanley had such tough skin but we would probably not assume that we would all be as impervious to knife wounds as Uncle Stanley. Even though inductive reasoning would lead us to conclude that sharp knives are bad for our skin, we would not seize upon the exception provided by Uncle Stanley to conclude that the basic premise of this inductive reasoning process was flawed. Certainly we would be more likely to conclude that Uncle Stanley is the rare exception to the rule and that we should not plunge knives into our bodies even though Uncle Stanley might call us "woosies."

So reasoning by induction is not foolproof and we may indeed end up with an unexpected (and incorrect) result. But surely there must be another way we can avoid the pitfalls of reasoning by analogy and inductive reasoning—or at least a reasoning process in which there is a greater assurance that the conclusion drawn from the facts observed will be logically and factually consistent. Fortunately for humanity, there is a wonderful reasoning process known as deductive reasoning whereby even the dimmest of nitwits can construct an airtight argument based upon the assertions of the basic premises. What is particularly nice about deductive reasoning, as we shall see shortly, is that it avoids the trap that is possible in inductive reasoning: the inability to ever arrive at a completely conclusive result. What do we mean by this statement? We already know that inductive reasoning entails the drawing of certain conclusions based upon repeated observations. But let us look at this problem another way. Suppose you are the world's foremost expert on clover and you edit the prestigious *Journal of Clover* which is regarded by its eight subscribers as the leading scientific journal dealing with the subject. As you have spent the better part of your professional career crawling around on all fours through meadows in search of unique species of clover, you might regard yourself as having a pretty clear idea as to the range of physical characteristics in the world's clover. By this time in your career, in fact, you might have concluded that there is no such thing as a yellow four-leaf clover even though two of your subscribers reported seeing one. Even though you might be skeptical about the claims of your colleagues, you would have no way, based upon inductive reasoning, to prove conclusively that there is no such thing as a yellow four-leaf clover unless you were able to cover every square foot of ground on the planet and verify that such a clover does not exist. This illustrates the trap which lies beneath the seemingly tranquil surface of inductive reasoning and that is the impossibility of proving a nonexistent fact. Now it would be quite another thing to try to prove the existence of a yellow four-leaf clover because once a specimen was found, then the search for the truth would end at that point. So every scientist avoids becoming involved in matters which essentially require the proving of facts which do not exist because they essentially give rise to open-ended searches which can never be authoritatively concluded.

Inductive reasoning, like reasoning by analogy, involves an intellectual process that is essentially the reverse of that which is needed to construct any type of mathematical edifice such as the calculus. After all, there is no observation in nature which can lead us to formulate the basic principles of integral or differential calculus. We can only hope that we will see certain phenomena such as the planets spinning around the sun or objects falling from the sky that will cause us to wonder whether there is some way that such dynamic actions can be expressed mathematically. But the formulation of a mathematics that can express rates of change requires a completely different type of reasoning process.

Now that we have disposed of both reasoning by analogy and inductive reasoning, we need to consider deductive reasoning. Every student who has sat through a course in logic will immediately appreciate the basis for deductive reasoning. Sometimes it will be rooted in an "if, then" statement such as the following: "If Eddie is a sanitation worker and all sanitation workers belong to country clubs, then Eddie belongs to a country club." Here we are not concerned with whether Eddie is an equity member or whether he pays his dues on time or whether he wears plaid knickers on the golf course but merely with the logical consistency of the statement itself. We can determine very quickly by a quick sniff of the air whether Eddie is a sanitation worker. If all sanitation workers belong to country clubs, then we know without a doubt that Eddie belongs to a country club. The wonderful thing about deductive reasoning is that it is devoid of wishy-washiness. Once you accept the correctness of the initial assumptions, the conclusion logically follows from their premises.

Of course the strength of this form of reasoning is totally dependent on the correctness of the basic premises. If only a few sanitation workers belong to country clubs, for example, then the fact that Eddie is a sanitation worker does not force us to conclude that Eddie belongs to a country club. But the preceding examples illustrate how facts can be combined to obtain a new but equally valid fact. It also hints at the power that may be wielded through the use of deductive reasoning. The mathematician Morris Kline noted that "a deductive argument consists in combining accepted facts in any way that compels acceptance of the conclusion."* But Kline also points out that "this characterization of deductive reasoning does not specify just what kinds of combinations of accepted facts yield inescapable conclusions."† In other words, anyone who wants to use deductive reasoning will need to verify the validity of the premises themselves because junky premises will yield junky conclusions.

Such a form of reasoning would seem to be the only choice if indeed it is as certain and as uncompromising as that posed by deductive reasoning. One can imagine a legion of soldiers, each clad

*Morris Kline, *Mathematics and the Physical World.* New York: Dover, 1959, p. 15.
†Ibid.

in a uniform with a big "D" (for "Deductive") on their chests, moving inexorably across the plains, striking down enemy soldiers who owe their allegiance to the competing camps of either inductive reasoning or reasoning by analogy. This idea of using an *if, then* format to extract logically unimpeachable truths has a certain appeal to it, particularly when we must so often deal with vagueness and ambiguity in our everyday lives. But we also need to bear in mind that deductive reasoning, while forming the linchpin for nearly all that is important in mathematics (particularly calculus), is not the be-all and end-all of intellectual achievement. We can see why it would appeal to our desire for order and rationality because it does make it possible to create sequential thoughts. But there are situations in which deductive reasoning will not be very useful and one may indeed have to return to those discards—inductive reasoning and reasoning by analogy.

The value of deductive reasoning depends in large part on the validity of the premises which give rise to the conclusory statement. Let us take the following statement: "If George Washington was President of the United States and only American citizens can become President of the United States, then George Washington was an American citizen." Here, we can see that the premises appear to be valid and they in turn make it possible for us to draw a seemingly indisputable conclusion that only American citizens are eligible to serve as President. But the strength of this argument is only as great as its weakest link; if any single premise is not true or is true in only certain situations, then the force of the argument is compromised or wrecked entirely. If, for example, George Washington had not been President but instead had been a female impersonator or a wig model, then the entire logical coherence of the preceding statement would be undermined. It would make no sense to say the following: "If George Washington were a female impersonator and only American citizens can become President, then George Washington was an American citizen." We can see quite clearly that the conclusion does not automatically follow from the premises because the premises themselves do not really correlate with each other. The same result would hold true if the sentence were restated as follows: "If George Washington was President of the United States and only American citizens can bake fluffy cakes, then George Washington was an American citizen."

Although the validity of the premises will affect the confidence we may have in the conclusion of a deductive statement, we also need to realize that deductive reasoning is not an all-purpose type of reasoning. In particular, it is not terribly helpful when we are dealing with nonquantitative or subjective statements such as the following: "Chocolate ice cream is the yummiest flavor." Although some may quibble with the argument that the word "yummiest" does not have a strong quantitative aspect, it is difficult to imagine how one could construct a deductive argument that would have any meaning. In other words, we could try to prove that chocolate ice cream is the yummiest flavor but there are no obvious criteria that will invariably lead us to a deductive conclusion. Here we might instead use something akin to inductive reasoning and try many different flavors of ice cream ranging from spam and chocolate fudge swirl to anthrax sherbet before concluding that chocolate was indeed the yummiest flavor. This would not be a completely satisfactory experiment because we would still lack a quantitative standard but it would be a kind of battery of repeated taste tests which would ultimately lead the investigator to indicate a preference for a particular flavor—that which we would consider the "yummiest" flavor of all.

Our previous discussion of reasoning by analogy implicitly suggested that it was not terribly useful for scientific work. But reasoning by analogy can come into play, particularly in areas involving medical experiments. Nearly every drug manufactured today is tested on rats or other small mammals first before being administered to human beings. The reason for this testing sequence is obvious: Rats are much easier to pin to the floor and inject with a syringe than are full-grown human beings. They are also much less likely to contact personal-injury lawyers and sue for damages. Still another reason is that rats and other small, warm-blooded mammals have certain physiological features which are similar to those of humans so that the effects of certain drugs administered to rats, for example, may be expected to manifest in similar ways in humans. Once we find that a certain drug called Agrax causes rats to grow bushy eyebrows, for example, then we will wonder whether the drug will do the same for humans who desire to have the fresh-scrubbed look of those old-style Soviet dictators. We will be engaged in a process of reasoning by analogy if we try to determine which features common to the physiologies of rats and human beings will make it possible for the Agrax to have a similarly beneficial effect on eyebrowless human beings. We will also try to figure out what sort of differences between the two physiologies might prevent the Agrax from having any perceptible effect on humans. Of course the proof is in the pudding (along with the cook's fingernail), and reasoning by analogy will only carry us so far. As a result, we will not, in most cases, be able to say with any certainty that Agrax will work on humans until we actually try it on humans. But we will have accumulated enough knowledge during our prior experiments to know how best to minimize any adverse consequences it might have on its human subjects. However, there is always an element of uncertainty involved in such tests and our reasoning by analogy, no matter how well thought out, may simply be wrong. It may be that a tiny dose of Agrax kills each of our human subjects, thus leaving us with a rather embarrassing situation to explain to the press.

Inductive reasoning may also be used as a sort of trial by error reasoning process by one who may not be familiar enough with the premises of a problem to use deductive reasoning. Suppose that Elroy has just finished a six-pack of beer and is feeling so creative that he invents the entire field of plane geometry on a napkin while sitting in front of the television set in his mobile home watching professional wrestling. As Elroy has not yet learned that his masterpiece was preceded some 2000 years before by Euclid's 13-volume *Elements*, he will be feeling quite pleased with himself. But the approaches used by Euclid and Elroy are different. Whereas Euclid based his geometry upon the construction of certain fundamental postulates from which many conclusions could be drawn, Elroy employed inductive reasoning and a great deal of guesswork in fashioning his own "Elroyian" geometry. The difference in the two approaches can be illustrated in the way that each of them determined that only one line passing through a point can be drawn perpendicular to a straight line. Euclid, in constructing his postulates, would have deduced that such a conclusion invariably follows from the basic premises of his geometry, more specifically, the fact that a straight line may be extended to infinity. Although one could have an infinite number of lines intersecting this straight line through a given point on that line, there is only one line that can pass through it at a 90-degree angle (perpendicular to the first line). Elroy, on the other hand, might pull out a ruler and a pencil and begin trying to figure out how many lines could pass through a second line at a 90-degree angle. He could sharpen his pencil and get a fresh napkin and draw the straightest possible lines he could draw through a second line until the paper began to tear. After a while Elroy might decide that there did not seem to be enough room for very many lines to pass perpendicular (90 degrees) to the original line. A few more napkins and a break watching a "steel cage match" on television might enable Elroy to clear his mind and realize that there probably is only one line that can intersect at a 90-degree angle with another line through a given point. This revelation would bring utter joy to Elroy's heart and he would celebrate his discovery by emptying a round of buckshot into his ceiling.

Although Elroy's method may have some appeal, it is not one that has found favor with mathematicians. "Despite the advantages that may accrue from other methods of obtaining knowledge, mathematicians ever since Greek times have limited themselves to conclusions which can be established deductively on the basis of a fixed set of thoroughly reliable premises."* The growth of mathematics in general has been made possible by the use of the deductive method to derive certain conclusions which in turn may serve as premises for the drawing of new conclusions. So mathematicians have proceeded from the beginning to build the entire body of mathematics from the ground up like a never-ending staircase, with each new step totally dependent on the support provided by the previous step. Mathematicians have thus strived to maintain the logical consistency and intellectual purity of mathematics by resorting purely to deductive reasoning in the formulation of new principles. Of course the risk of this approach is that there is some fundamental flaw in the foundation of this elaborate edifice which has escaped detection up to now

*Kline, op. cit., p. 17.

and would, if found, bring the entire structure tumbling down. Admittedly, this risk is quite remote due to the nature of the deductive reasoning process itself which relies solely on the validity of its premises for its conclusions. But it would be a sad day if we found out that Euclid, for example, was a horrible speller and suffered from bouts of lunacy and that his *Elements* was actually a very technical marital aid book. Of course a proper translation of Euclid's work might cause interest in what we formerly knew as Euclidean geometry to skyrocket so much that people who never even showed the slightest interest in geometry would be picking up Euclid's book and reading all about his thoughts on the parallel postulate, a concept which would now be shown to have more to do with the missionary position and axle grease than with lines and points.

But the chances of finding such a flaw in the foundation of mathematics is quite small because of the very nature of deductive reasoning. Admittedly, all mathematics begins with certain arbitrary assumptions because it is, after all, an intellectual construct that does not have a true counterpart in the physical universe. As such, one must begin by making certain arbitrary statements to at least have some way of going forward. There is nothing magical about the assumption that only one line can pass through another line at a perpendicular (90-degree) angle through a single point in a plane. There is no natural phenomenon that forces us to draw such a conclusion. We do not see atomic particles moving around in such a way to compel us to accept this idea nor is there anything in the structure of the DNA molecule that validates this postulate. Instead we accept it as being true based upon the reasoning process we use which we know is totally dependent on the means by which we draw conclusions from basic ideas.

Although one can quibble with the absence in mathematics of a real-world basis, it is certainly true that mathematics has acquired a reputation for certainty and precision that is the envy of all fields of knowledge. This reputation has developed precisely because mathematics is not dependent on the messiness of the real world for validation. Nothing is so pristine as a logical thought. It certainly is a simpler beast than the often chaotic displays of physical phenomena which we find in the real world such as the seemingly unending arrays of subatomic particles which seem to defy a unifying order or the behaviors of large numbers of living organisms. But it is this very separation from the real world which prompts the critics of mathematics (particularly grumpy students who are wading through the basic principles of algebra) to criticize mathematics as being irrelevant to their everyday lives.

Unlike mathematicians, most scientists use a mix of deductive reasoning, inductive reasoning, and reasoning by analogy. This multi-pronged approach has nothing to do with scientists being "swingers" or having a greater willingness to try exotic foods. Instead it stems from the fact that scientists have as their laboratory the entire universe and must necessarily deal with the myriad phenomena contained within that universe. They cannot focus on purely intellectual constructs which do not readily relate to the real world because they would thus be precluded from making meaningful statements about the real world. A scientist who studies the chemical reactions that occur when two substances are mixed together, for example, cannot learn everything that is possible to learn about that mixture if he or she never bothers to actually examine what happens when the two substances are combined. It would be a little like Pythagoras drawing upon his knowledge of triangles to hypothesize that one can build a particular type of bridge using a triangular support system without testing any materials for strength or making any other type of preliminary investigation. We would not take Pythagoras seriously because he would be relying on deductive reasoning to answer a problem that cannot be solved by purely deductive means—whether a particular material such as peanut butter or molasses can be used to construct a bridge.

Because scientists must grapple with the phenomena of the real world, they cannot limit themselves to deductive reasoning. Even Albert Einstein, who, along with Isaac Newton, was arguably one of the two most important scientists who ever lived, did not restrict himself to purely deductive reasoning. Einstein, for example, utilized "thought" experiments, particularly in his general theory of relativity, to try to understand the effects of the gravitational force on a body. In particular, he would imagine what would happen to a person standing in an elevator as it was accelerated toward a massive body to better understand the manifestations of that body's gravitational field. Einstein also proposed the so-called "twin paradox" in his special theory of relativity which showed how the passage of time for a twin who traveled on a spaceship at a velocity near the speed of light would be slower than that for the other twin who had remained behind on earth. These thought experiments enabled Einstein to visualize how the principles of his relativity theory would manifest physically. These mental experiments thus made it possible for him to see (1) how the gravitational field generated by all the objects in the universe, for example, could determine the ultimate fate of the universe itself or (2) how his famous mass–energy equivalence equation $E = mc^2$ would make possible the atomic bomb due to the vast amounts of energy which could be released by destroying even very small amounts of matter.

But Einstein did not neglect the use of deductive reasoning in his work even though he found much could be learned by reasoning by analogy. In particular, Einstein's theoretical work was dependent in large part on the mathematical work of Georg Riemann, whom we discussed previously, and William Kingdom Clifford, two 19th-century mathematicians who had done extensive work in the mathematics of curved space. As you will recall, Riemann, in particular, had revised Euclid's famous parallel postulate to create a geometry of positively curved surfaces. The so-called Riemannian geometry provided much of the theoretical foundation for Einstein's general theory of relativity, which proposed that space would be curved by the gravitational fields of massive objects. So Einstein was able to build upon the deductive reasoning used by Riemann to create his geometry and then fold Riemann's work into his own theories about the nature of space, time, and gravity to create his general theory of relativity. So, in a way, Einstein used both deductive reasoning and reasoning by analogy in his work. However, much of the deductive reasoning had already been carried out by these prescient 19th-century mathematicians.

Certainly there is an argument to be made for having deductive reasoning, inductive reasoning, and reasoning by analogy in one's bag of tricks. And one can go out and join a group of picketers carrying signs which accuse mathematicians of being too narrow-minded because they rely solely on deductive reasoning. But you might find it a very unimpressive rally because not too many people really care a whit about whether mathematicians should use inductive reasoning or reasoning by analogy in considering mathematical problems.

The mathematician's reliance on deductive reasoning is not totally misplaced because it does make it possible for the realm of mathematical thought to be completely self-contained. In other words, the theoretical foundations of mathematics are purely conceptual in nature and do not need to be reconciled with an imperfect physical world. The cohesion and unity of mathematics is based on the process by which postulates are formulated and then used to draw conclusions from which further postulates can be formulated. In the physical sciences, by contrast, the physicist or the biologist must always bear in mind that nature will not necessarily go along with every theoretical prediction. As the universe is only partially understood, the fact that a scientist may propose a logically consistent theory does not guarantee that the theory itself will be supported by observation.

So the mathematician is limited only by the scope of his imagination and the logical constraints of his mathematics. Indeed, physicists or astronomers are greatly limited by comparison because they must constantly check their theories against their observations. Mathematicians, by contrast, can allow their imaginations to run wild so long as they observe the basic rules of organization and construction regarding their subject. A mathematician, for example, can easily describe a ten-dimensional space even though the physicist has a very difficult time imagining how such a universe could exist or even be understood. Similarly, mathematicians think nothing of playing with the concept of infinity and could easily borrow upon the 19th-century German mathematician Georg Cantor's unending hierarchy of infinities (the so-called transfinite numbers which are used to represent the cardinalities of infinite numbers, which we shall discuss in a later chapter). At the risk of jumping the gun, a mathematician could take the smallest transfinite number, aleph-null, and assign it arbitrarily to represent the cardinality of our infinite three-dimensional universe. Hence, the mathematician would be taking a physical infinity (a presumably unending universe) and representing it with the mathematical term which describes the most basic infinite quantity. Our mathematician would then proceed from there to create so-called transfinite universes which are basically obtained by raising the cardinality of the base infinity (aleph-null) assigned to our own universe to its own exponent (aleph-null to the aleph-null power) to create a transfinite universe having a cardinality of aleph-one. Now nobody would have the slightest idea as to what a transfinite universe with a cardinality of aleph-one would look like but it would exist in a mathematical sense. Despite the strangeness of this idea, the cardinality of this transfinite universe could in turn be raised by its own exponent (aleph-one) to create a new transfinite universe having a cardinality of aleph-two and this process could be continued without limit. By carrying out simple mathematical operations, the mathematician would thus be able to create a type of mathematical universe which is unimaginably vaster and infinitely "denser" in a mathematical sense than the infinite ocean of space in which our own universe floats. But the physicist would have little use for such a grand vision because he would have no way of actually verifying the existence of these transfinite universes. He would either have to accept the mathematical model on faith (as has happened on more than one occasion in the history of physics) or simply ignore its ramifications as nothing more than a clever but irrelevant mathematical trick. Indeed, the physicist would be at a loss to understand how he or she could describe a transfinite universe when his or her own experiences are confined to a comparatively simple three-dimensional universe having the base cardinality (arbitrarily assigned) of aleph-null. But the mathematician could say that the transfinite mathematics provides a blueprint of sorts which offers something of an identification process.

The physicist, then, is necessarily limiting himself to the observable universe because it is the world of stars and atoms which provides the physicist with the laboratory in which the predictions of theories may be tested. The mathematician Morris Kline, for one, views the physical scientist's reliance on real-world observation and perception as a distinct disadvantage:

The senses are limited. The eye sees only a small range of light waves and is easily deceived as to sizes and locations of objects. The ear hears only a limited range of sound waves. The sense of touch can reach only to objects accessible to the hand and is not a precise sense at that. On the other hand, man's reason can encompass distances, sizes, sounds, and temperatures beyond the range of the senses. More than that, reason can contemplate phenomena which transcend the senses and even the imagination. Mathematics has thereby been able to create spaces of arbitrary dimension and to predict the existence of imperceptible radio waves. And because mathematics has confined itself to the soundest methods of reasoning man has, the results of mathematics have endured whereas even some of the most magnificent theories of science have had to be discarded.*

Even the most hardbitten physicist would have to agree with Kline's assessment, which is very compelling due to the power of deductive reasoning. But by the same token, mathematicians must confine themselves to arenas in which fruitful deductive reasoning is actually possible.

Indeed, one can prove many things to be true based on deductive reasoning but the analyses may not necessarily lead the investigator to any meaningful conclusions. If we want to prove that most people in the world have ten toenails, then we would formulate premises which state that most people are born with ten toes and that each toe of each person typically has one toenail. This is not the type of far-ranging investigation which causes members of tenure committees to swoon with delight but it does remind us of the invariable way in which conclusions will follow from premises. No doubt our reasoning regarding toenails will be aided by our own firsthand experiences on public beaches and at public pools where we will have observed most people to have ten toenails (except for Great Uncle Albert who had worked as a janitor near a nuclear test site in the Nevada desert and can now hang from a tree with his sixteen toes). But these premises would lead us to the conclusion that most people do, in fact, have ten toenails. Unfortunately, this grand revela-

*Ibid., p. 19.

tion would not really lead us to any profound conclusions because there is no insight to be gained, no dramatic conclusion to be revealed by determining that most people have ten toenails.

But is deductive reasoning as pure and as distinct from the real world as it would first appear to be? Although mathematics is based on logical consistency, it seems farfetched to suppose that all of our mathematics simply sprang full-grown from the minds of several brilliant mathematicians over a period of some 2000 years. Indeed, one would wonder how mathematics could have any relevance to the real world if it was not inspired in part by the manifestations of nature itself. Nearly every simple geometric shape can be found in nature—whether it be the triangular mountains or the circular moon or the cylindrical trees. It is not clear who was the first person to distill the essential shape of a triangle or a square from a natural object but it is difficult to imagine why anyone would have bothered to create these conceptual shapes out of thin air. After all, these shapes would seem to have no more value than random doodles on a sheet of paper if there did not seem to be some type of correspondence with nature's objects. But there is obviously something more at work in the process of mathematical abstraction than merely drawing simple figures. Trial and error and dumb luck do not get us from sketching line drawings of natural objects to rigorous mathematical proofs. Instead there is some sort of separate mental process at work here that requires mathematicians to engage in a creative process of abstraction.

Not only must mathematicians rely on the deductive process for constructing their mathematics, but they must also use abstract concepts. By and large, these abstract concepts are drawn from the properties of physical objects. A perfectly round stone, for example, is a sphere and the mathematician looking at it would notice that it had a number of physical properties including weight, texture, color, and hardness. A sphere drawn by a mathematician, however, would have none of these features. It would be little more than a pictorial depiction of the abstract concept of a perfectly round, three-dimensional object.

But the process of abstraction does not end with simple physical objects and mathematicians have felt free to create an entire gallery of strange, even bizarre, numbers including negative numbers and irrational numbers. Mathematicians have also created other concepts which do not resemble anything in the real world including formulas, derivatives, and integrals, which are even further removed from the physical world because they can be derived only from other abstract concepts such as numbers. In a sense, then, this is where mathematics becomes even more susceptible to being challenged as not being part of the real world.

But the process of abstraction is not so unreal or irrelevant as some would charge. We all engage in the process of abstraction when we solve simple problems like deciding which route is the shortest to a particular destination or which of several competing brands of cat litter we should purchase as a dinner appetizer for an upcoming visit by the in-laws. Most of us also daydream about the ideal man or woman and such idle thoughts may lead us to create in our own minds a person who does not actually exist, at least in the absence of plastic surgery. This process of abstraction can be pleasurable and certainly less contentious than a relationship with an actual person.

But the value of abstraction lies in the simplification of an object to its bare essence or its basic features so that it can be studied. Of course mathematics is not the only discipline which uses abstractions to gain a greater understanding of its subject matter. But it is fair to say that the abstractions that are used by mathematicians are less tangible than those concepts used by scholars in most other fields. A physicist, for example, uses concepts such as force, mass, and momentum to better understand the manifestations of forces in nature. Similarly, a psychologist will grapple with such concepts as id, ego, and superego to gain greater insights into the mental state of his or her patient. But these concepts are used to deal directly with material objects. The physicist, for example, is trying to use these concepts to explain how matter and energy interact in the real world. Similarly, the psychologist is using these concepts to understand why his patient continues to try to bludgeon the mail carrier with an axe every time he delivers a piece of junk mail. These terms can also sound very impressive to the average patient and help to convince that patient of the need to continue many years of expensive but helpful therapy.

But the reliance of mathematics on abstraction appears to be more pronounced than that of any other discipline. This statement must be qualified somewhat because so many of the social sciences, such as economics and political science, rely heavily on mathematics to construct sophisticated models. Whether these models actually have any relevance or, indeed, any point is a judgment best left to the individual to make, but professional scientists in these disciplines would argue that mathematics serves a very important purpose of conveying basic concepts or ideas with a minimal number of symbols. In other words, a single symbol may be used by an economist, for example, to express a complex concept such as the elasticity of demand for a product, which refers to the extent to which a given increase in the price of a product will reduce the consumption of that product. So the elasticity of demand reflects the extent to which we can forego consumption of that product if its price increases significantly. An increase in the price of gasoline, for example, will not affect consumption significantly because there are no perfect substitutes for gasoline as anyone who has ever tried to fill his gas tank with water will attest. But an increase in the price of orange juice may cause a great decline in the amount of orange juice consumed because there are so many substitutes for orange juice which may be set on the morning breakfast table including milk, coffee, apple juice, grapefruit juice, and straight vodka. As most economists are swinging people on the go with lots of important things to do, they do not want to have to explain the concept of elasticity of demand every time they pick up a pencil to write a paper for publication. As a result, they will use a mathematical symbol to represent this concept. In this way, the symbol becomes the concept itself. If you ever go to a convention of economists, you will find that some of the most honored economists are those who have been most successful in reducing the most complex ideas to a series of mathematical expressions.

The social sciences deal with real things such as the flow of money in the economy or the behavior of individuals in primitive civilizations. But they also depend on mathematics for much of their analytical usefulness. As a result, one cannot neatly characterize them as relying solely on inductive reasoning or reasoning by analogy or deductive reasoning. But the fact that so much of the social sciences is based on mathematical modeling does mean that the individual scientist must be somewhat skilled in the art of deductive reasoning and be able to recognize the flaws that can arise in the reasoning process. Yet it can also have a liberating effect because a singular reliance on deductive reasoning can lift one's intellect out of the gutter of prosaic detail in which many physical scientists are forced to wallow. The mathematical physicist, for example, does not have to accelerate particles he cannot see into other particles he cannot see to discover ever-smaller constituent particles because he is able to use mathematical models to predict the behavior of atomic particles. The actual experiments are left to those research physicists who enjoy rolling up their sleeves and playing with big shiny atom smashers. The mathematical physicist, however, fancies himself as being at the top of the physicist hierarchy because his work most closely approximates that of the pure mathematician. He does not need an elaborate laboratory for testing the viscosity of substances or the strength of the earth's electromagnetic field because his work can be carried out on a chalkboard or a calculator or a supercomputer. His work is as abstract as that of the mathematician and is thus as free of the constraints of inductive reasoning or reasoning by analogy as any branch of science can be.

The use of deductive reasoning is akin to flying over the landscape on which most scientists must travel. One ignores the details of the physical world so as to be able to concentrate on a few properties which are of interest to the scientist. In a sense, the use of deductive reasoning requires one to ignore the details of individual trees in favor of the forest itself. A scientist seeking to learn more about the mechanics of the solar system would try to express the observed orbits of the planets in the most general form possible; he would not try to calculate the movements of every planet at every possible point. Indeed, this was the approach adopted by the German astronomer Johannes Kepler who devised three laws to describe the motions of the planets in the solar system. Although Kepler was able to check the accuracy of his laws of planetary motion against the tables of observations made by the Danish naked-eye astronomer Tycho Brahe, Kepler's laws, particularly his third law (which states that the square of the orbital period of each planet is proportional to the cube of its mean distance from the sun), were dependent on deductive reasoning. But the true beauty of Kepler's work, which illustrates the grandeur of deductive reasoning, is its universal application. His first law, which proposes that the planets move in elliptical orbits around the sun, is applicable to every planetary system in the universe so that Kepler's work provides a model for the billion trillion planetary systems which may exist in our universe. This is certainly a much more elegant (and time-efficient) method than studying every individual system—a supreme illustration of the advantage of the deductive as opposed to the inductive approach.

The natural consequence of abstracting physical phenomena is that it entails the filtering out of the details of the problems which are to be solved. If the Italian astronomer Galileo Galilei were interested in tossing objects off the top of the Leaning Tower of Pisa to see whether the weight of a given object affects the speed at which it falls through the air to the ground, his notebook would not be laden with details about the blue-green swirls on the surface of his bowling ball or the comforting smoothness of his satin sheets as he heaved them over the side of the railing. Instead he would probably limit himself to describing the item by name and weight and then recording the time it took each object to fall from the top of the tower to the ground. Over time Galileo would find that it did not matter whether he threw a feather pillow or a lawnmower or even an inflatable raft over the side, as all of the objects would fall at the same rate of speed to the ground (if the air resistance were ignored). If Galileo could find some dimwitted volunteers, he would find that this relationship also holds true for humans--whether they be thin or fat or even pleasingly plump. But here again Galileo would be interested in noting only the barest of details about each individual person who he managed to push over the side such as their weight and, perhaps, the amount of time they screamed before they hit the ground.

It is the process of abstraction and idealization that makes it possible for the full power of deductive reasoning to be realized in both mathematics and the sciences. Abstraction makes it possible for the mathematician to focus on the essential features of a problem or for the physicist to concentrate on isolating the important physical characteristics of an experiment. It thus makes possible a process of both simplification and generalization which has propelled our body of knowledge forward from the dark ignorance of antiquity to the mind-boggling advances which have made possible our own technologically advanced society. It is also the process which has made it possible for mathematics to advance beyond the static progeny of Euclid and Archimedes to the dynamic mathematics offered by Newton which could quantify rates of change and thus encapsulate the motions of a complex universe in a series of concise equations. Thus the discovery of the calculus could be said to represent a sort of culmination in the evolution of mathematical thought, much in the same way that relativity theory and quantum theory can be viewed as the crowning glory of modern physical thought. Although the calculus and relativistic physics claim different realms of knowledge, both are vitally dependent on the logical underpinnings of deductive reasoning and, in this very fundamental sense, are inseparable.

CHAPTER 3

Groping toward Arithmetic

Except the blind forces of nature, nothing moves in this world which is not Greek in origin.
—Sir Henry James Sumner Maine

Every mathematics teacher knows that there is no better way to bring shouts of joy to a class than to announce the start of a new lesson in arithmetic. Yes, you can almost see the tears of joy in the students' eyes when they hear that their knowledge of this most basic branch of mathematics will be enhanced, whether it be by a drill or by a lecture or even by a sound beating. But such is the sad state of public relations for most of mathematics and arithmetic, the most fundamental part of mathematics. In order for a student to have any hope of mastering the branches of higher mathematics such as the calculus, he or she must first successfully navigate the shallow waters of arithmetic. Arithmetic—not algebra or geometry or the calculus—is the mother lode of mathematical reasoning as well as the foundation from which all mathematics arises.

But the hostility that many people have toward mathematics in general and arithmetic in particular seems to stem from the general perception that it is repetitious and boring. Of course many adult couples engage in marital activities which warrant the same criticisms but you will not typically find them proposing a complete cessation of these activities merely because they are dull or have no relevance to the workplace. Yet arithmetic remains a popular target, perhaps because it is such a fundamental part of our knowledge about the physical world and requires a certain degree of persistence to master beyond its most elementary aspects. We humans do not like regimentation and the pursuit of mathematics necessitates that one engage in systematic and, to some extent, regimented intellectual exercises. This sort of demand strikes many of today's more free-spirited and self-absorbed students as "lame" and not befitting the substantial amount of time needed to attain a thorough knowledge of the subject.

The Legacy of Pythagoras

Although bits and pieces of what we now know as arithmetic are littered throughout the wreckages of many ancient civilizations, most historians credit the early Greeks with having made the first important discoveries about what they called *arithmetica*. For the Greeks, arithmetic was more than a few rules or principles; it instead entailed a deeper understanding of the relationship between abstract numbers and the surrounding physical world. The Pythagoreans, in particular, despite their reputation for being part of a cult which embraced the ideas of women's rights and free love, were obsessed with numbers and offered a philosophy drenched in mysticism and numerology. Their founder, Pythagoras, was born about 569 B.C. in Samos. The adult Pythagoras migrated to the Greek colony of Croton on the southeastern coast of Italy where he founded a school. His dynamic personality soon attracted a loyal following which included women as well as men (which doubtless made the lab work on the free love lectures more realistic). The topics of discussion ranged from religion and philosophy to science and mathematics. Indeed, Pythagoras, like most of his contemporaries, did not worry about drawing distinctions between these various disciplines as they were all viewed as being different aspects of a single body of knowledge. This somewhat murky approach toward knowledge and learning did not facilitate the type of rational inquiry to which we are accustomed today but Pythagoras did offer a number of important insights about numbers and arithmetic which remain with us to the present. However, the Pythagorean school itself was not so enduring.

Its secretive ways aroused the suspicions of the Crotons, who, having tired of engaging in malicious gossip, finally forced the Pythagoreans to leave. Pythagoras himself was separated from many of his followers and eventually made his way to Metapontum where he might have spent a long and happy life surrounded by lovely maidens dropping grapes into his mouth were it not for the unfortunate timing of his murder there. The identity of his killer or the motives behind the slaying have never been conclusively demonstrated and the lack of available eyewitnesses suggests that this mystery will never be solved. But it is probable that Pythagoras's radical views on everything from women's rights to human sexuality aroused the anger of the local population and led to his murder. Pythagoras's followers then made their way to other cities throughout Asia Minor and thus perpetuated Pythagoras's legacy.

If we were to summarize the teachings of Pythagoras, we might declare: "All is number." This sounds either like a very profound thought or like the rants of a village idiot. But the Pythagoreans were very serious about the idea that numbers are the fundamental feature of the universe. They even assigned meanings to certain numbers, equating the number 1 with reason, the number 2 with diversity, the number 4 with justice, and so on. But their philosophy involved more than simply assigning vague meanings to individual numbers. The very substance of the universe was thought to consist of numbers and the four basic dimensions of geometry including points, lines, planes, and spaces constituted all of the possible physical shapes in which material objects could manifest. The Pythagoreans also believed that the universe consists of four basic elements—earth, air, fire, and water. But this conviction was prompted more by their addiction to the number concept than by any type of investigation into the nature of matter itself. The Pythagoreans also discovered that they could make harmonious sounds when two taut strings with lengths having the ratios of whole numbers were plucked. In this way, they discovered the musical concept of the octave and thus provided inspiration for future generations of musicians, from classical pianists to heavy metal guitarists.

Although the number 4 was obviously a sentimental favorite of the Pythagoreans, one only had to mention the number 10 to drive the typical Pythagorean into fits of ecstasy. The number 10 was obtained by adding up 1, 2, 3, and 4. This in itself was enough to convince the Pythagoreans that there must be ten heavenly bodies moving through space. Mind you, the Pythagoreans only knew of the moon, the sun, and a few nearby planets. But the number 10 was the number 10 so one could hardly dispute the fearless logic used by the Pythagoreans to construct their solar system. They reserved for 10 itself the honor as the number of the universe because it included the sum of all possible dimensions: "A single point is the generator of dimensions, two points determine a line of dimension one, three points (not on a line) determine a triangle with area of dimension two, and four points (not in a plane) determine a tetrahedron with volume of dimension three; the sum of the numbers representing all dimensions therefore is the revered number ten."[*]

Despite their tendency to look for a numerological basis in everything around them, the Pythagoreans were not vapid mystics who wandered around muttering nonsensical phrases about numbers. Instead they lived a fairly ordered life, embracing a conservative political philosophy and refraining from the consumption of meats. As much of their studies delved into religion, the Pythagoreans were greatly concerned with the idea of transmigrating souls. They also believed that their studies in mathematics could provide an ethical basis for the conduct of their daily affairs.[†] The Pythagoreans also banned the consumption of beans, a restriction that may have been prompted more by their cramped indoor quarters than any compelling ideological reasons.

Despite their somewhat quirky diets, however, the Pythagoreans do deserve a great deal of credit for helping to organize the study of what was then a very fragmented body of mathematical knowledge. In most earlier civilizations such as Babylonia and Egypt, the use of arithmetic had been fairly straightforward in that numbers were used to represent quantities of objects and nothing more. The Pythagoreans, by contrast, sought to construct an entire intellectual edifice for mathematics, reflecting a love of wisdom that

---

[*]Carl B. Boyer, *A History of Mathematics*. Princeton: Princeton University Press, 1985, p. 58.
[†]Ibid., p. 53.

would forever change the role of mathematics in society.* Their discovery of the harmonic ratios of plucked strings, for example, prompted the Pythagoreans to extrapolate this finding to the celestial bodies. As a result, they decided that the heavenly bodies must emit harmonic tones as they moved through space—the so-called harmony of the spheres. Whether this idea originated with Pythagoras or his followers is unclear. Yet it reflects both the daring as well as the incredible foolishness that crept into the Pythagorean theology, which in turn resulted in the chaotic mixing of fact and unbridled speculation. But at the same time it was difficult not to admire the sheer gallantry which Pythagoras and his followers must have possessed in order to complete many of their basic discoveries—including the Pythagorean theorem which holds that the sum of the squared sides of the two perpendicular sides of a right triangle is equal to the square of the third side, as shown in Figure 3. This is not to say that Pythagoras risked death for daring to scratch out the formula for the Pythagorean theorem in the sand because few passers-by would have bothered to inquire as to the nature of the scrawled symbols or cared to hear a labored explanation of the same. But it required a certain steely nerve to devote so much time to a project which many of Pythagoras's more "practical" contemporaries must have dismissed as a waste of time.

The extent to which Pythagoras was able to create a sophisticated, intellectual mathematics, which was so much closer to that of our own era than that of the more primitive, neighboring civilizations, has prompted many to doubt whether Pythagoras could have possibly wrenched mathematics from the simplistic numbering schemes of the Babylonians and Egyptians to the more abstract mathematics of today. Indeed, the later followers of Pythagoras have been suspected of having rewritten the history of Greek mathematics to place Pythagoras in a more favorable light. Due to the lack of extant written records about the mathematics used in the other Greek city-states, it is unclear whether Pythagorean mathematics was much more advanced than the mathematics used elsewhere in Greece at that time or if its written records were merely better preserved. But it

*Ibid.

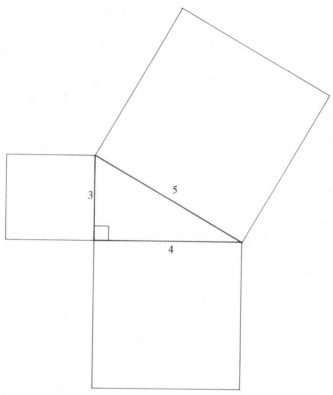

Figure 3. According to the Pythagorean Theorem, the sum of the smaller two squared sides of the right triangle is equal to the square of the hypotenuse (the longest side). In the above drawing, $3^2 + 4^2 = 5^2$. This statement is shown to be true when we see that $3^2 = 9$, $4^2 = 16$, and $5^2 = 25$.

is also true that Pythagoras did have a uniquely urbane attitude toward mathematics which would not have been found among the traders and merchants of that time. Certainly these businessmen were oriented toward tangible results and probably were not greatly interested in the deeper metaphysical questions about the relationship of mathematics to the real world. They also presumably did not have the leisure time nor the inclination to contemplate the philosophical implications of number theory.

Although it is easy to take jabs at the Pythagoreans for their mystical lapses or their dietary restrictions, they also deserve some credit for their belief that everything in the universe could be expressed in a numerical or quantifiable form. Indeed, the modern physicist can claim to be a kindred spirit with the Pythagoreans because numbers can be used to describe many features of the universe including the masses of its planets and moons and the energies locked in its stars. Certainly the physicist would appreciate the fact that the Pythagoreans saw numbers as a common thread underlying the chaotic landscape of the universe and thus offered a sort of unifying link to tie together things that would ordinarily appear to be completely dissimilar.

But the Pythagoreans also recognized that number is an abstraction separate and apart from the reality upon which it is superimposed. There is the concept of "oneness" and there is the single object which we describe numerically as "one." When dealing with actual objects, we see that "number is an abstract concept and that when numbers arise in physical situations, one must abstract the quantitative facts, perform the appropriate mathematical operations, and then interpret the mathematical answer."* Numbers are thus intellectual creations which exist separate and apart from physical objects but which are necessary for imposing order upon these objects and thus drawing meaningful conclusions about the arrangements of these objects.

We can better see the problems that arise from a failure to distinguish between abstraction and physical object if we follow Josiah to the local grocer. Josiah is a bouncy individual with a sunny disposition who has a kind word for everyone and he always looks forward to his trips to the grocery store. Not only can he stock up on his favorite brands of alcohol-based cough syrup and airplane glue for those days in which reality is a little too grating, he can also wave hello to the butcher who whistles happy tunes as he plunges his carving knife into beef hindquarters dangling from meathooks or flirt with Zelda, the sultry cashier, who is affectionately referred to as "the Maneater" by those who know her best. Anyway, Josiah ambles

*Morris Kline, *Mathematics and the Physical World*. New York: Dover, 1959, p. 31.

down the aisles stocking up on puffed rice and milk and eggs before going to the checkout counter to stare nervously at Zelda's pouty lips. Upon finding that the total bill is $31.00, Josiah will hand Zelda the appropriate amount of money and wait for his change and, perhaps, a hastily scrawled phone number. Although the transaction involves the payment of money for goods, we tend to forget that the interplay of abstract numbers and actual goods is a little more complex. In totaling the bill, Zelda is carrying out a purely mathematical operation because she is punching in the prices which have been affixed to each good. But the mathematical operation of adding up all of the prices requires Zelda to use numbers which in turn will lead her to a single grand total number. Once she has arrived at this number, she must then apply this result to the situation at hand. In effect, Zelda must apply the grand total of all of the numbers she punched into the cash register to the groceries piled on the counter. Because we do this type of operation every time we go shopping for food or clothes, this may not strike one as a particularly profound operation. But it is certainly the case that one cannot simply view numbers as a perfect reflection of reality. It would make no sense to speak of purchasing one-half of a pair of pants or two-thirds of a pet dog because, quite frankly, it would be pointless to wear one-half of a pair of pants to a formal dinner and rather gruesome to keep two-thirds of a dog at your home.

So one should always keep in mind the limitations inherent in using any abstraction—particularly numbers. Because we all have to consider the prices of individual items when we go shopping, we all carry out this process of abstraction, calculation, and interpretation whereby the abstract numbers are manipulated through the necessary mathematical operation (e.g., addition, subtraction, multiplication, division) and then the result is interpreted based upon its application to the actual situation being considered. So our purchase of several chic outfits by the famed designer Henri Revolteen from his controversial "Endangered Animals" line will require us to engage in a similar process of abstraction, calculation, and interpretation. As we must have the leopard coat with the white tiger skin lapels ($3000) as well as the rhino shoes with the whooping crane bill eyelets ($1400), we will have to take the prices for these two extremely tasteful pieces of merchandise, carry out the addition process, and then interpret the results. Here we would conclude that the total price of the two items is $4400. But before we can even draw this result we have to perform the purely mathematical operation of adding 3000 and 1400. It then becomes meaningful in terms of our purchases when we apply the sum of the two numbers to our purchases and thus determine that we may need to mortgage the house if we want to hang out with the beautiful people who wear Revolteen's line of clothing when they go out to the clubs.

You Ain't Seen Nothing Yet

Any discussion about arithmetic must necessarily include the number zero. Few things have caused more headaches for mathematicians and philosophers than the number zero because of the almost automatic tendency to equate the number zero with the concept of nothing. When we speak of zero, we are speaking of a quantitative amount, whereas when we refer to "nothing" we are talking about the complete absence of a particular physical object such as the author's lack of a fancy turbo-charged sports car. It makes sense to say that my ownership interest in a leech farm in Madagascar is nothing because I do not in fact have an investment in such an enterprise. If I did invest some money in this leech farm but a curious lack of demand for leeches in the world's culinary market caused the farm to go bankrupt, then my ownership interest would be worth zero dollars. "Nothing" is thus the ultimate in nihilistic concepts because it suggests a state that does not exist and did not exist in the past (my non-ownership interest in the leech farm), whereas "zero" suggests something akin to a state that exists but lacks an actual quantity. Similarly, if I am a chronic gambler who has been tossed out of various rehabilitation clinics due to my persistent recidivism (such as my repeated attempts to organize bookmaking activities among the other patients), then I can probably appreciate the distinction between nothing and zero. Due to a recent string of unfortunate bets on the three-legged elephant races in India, my total assets are equal to zero. This is a statement of quantity or, more properly, the absence of a given quantity. I have zero dollars in the bank and am now becoming much more open-minded about dating elderly, feeble-minded, wealthy widows. But the concept of nothing comes into play in a different way. If my assets never included shares of stock in the Flat Shoe Athletic Corporation, the only company which manufactures athletic footwear for morbidly obese track and field sprinters, then we could properly say that I have nothing in the way of shares of stock in that company. Had I once owned shares of stock in that company but lost them due to a unfortunate understanding between myself and a blackjack dealer at a table in Las Vegas, then we could say that my holdings of stock in that company are now equal to zero. The fact that I once held stock in that company would mean that there was a qualitative aspect to this statement which would not exist had I never held stock in the company at all.

Having firmed up our understanding of the difference between nothing and zero, we can now delve a little bit deeper into the concept of zero—specifically its unique properties as a whole number. Zero is special in that it occupies the center point of a number line which extends infinitely far in either direction. To the left of the zero mark on this number line are a series of negative whole numbers, beginning with -1, -2, -3, and continuing onward without limit at equally spaced intervals. To the right of the zero mark on this number line are a series of positive whole numbers, beginning with 1, 2, 3, and continuing onward without limit at equally spaced intervals. By virtue of its position as the linchpin of the number line, zero has certain mathematical properties which are unique to it. First of all, it starts with the letter "z." (Contrary to what many schoolchildren and adults might think, there is no such number as "zillion" even though it is a perfectly good name for a number with a lot of zeros and probably should be adopted by the leading mathematical associations and societies.) Second, the number zero cannot be divided by itself in any meaningful way. In other words, we can divide 0 by 0 which gives us the fraction $0/0$ which is undefined. But this does not necessarily help us very much because we find that we get a similar nonsimplistic answer if we divide 1 by 0 or if we divide 12 by 0 or any other number by 0 including 5,897,335,098,215,398. So we do not really get very far when we start reaching for long lists of numbers to divide by zero because every nonzero number divided by zero is equal to infinity. While this would make for a very nice mind-numbing activity that would cure even the most severely afflicted insomniacs, it would be pointless.

The development of zero was a crucial step in the continuing movement toward our modern system of numeration. The number zero made it possible to write any whole number of any desired magnitude. It also completed the development of our present base 10 numeration system. This system is based on the idea of grouping multiples of tens beginning with tens, tens of tens, and so on. That we have a base 10 system and ten fingers on our hands is probably not a mysterious coincidence. Although some persons might conclude that our base 10 notation was brought to humanity by those very same aliens who periodically visit the earth in their spaceships and conduct medical examinations on drunk hunters and hillbillies, it seems more likely that this was a product of humanity's own genius and a fortuitous number of digits. Fortunately for us, our base system was not set up by someone having 13 or 14 or even 15 fingers even though we would undoubtedly be as comfortable with such a system as we are with our own base 10 system. Of course a base 15 system would be easier to understand from an intuitive point of view if we all had 15 fingers.* But this of course would mean living in a world in which the bowling balls had four or five holes and hand gestures offered unprecedented opportunities for vulgar creativity and manicurists who charged their customers by the finger were among the wealthiest members of society. But our sturdy base 10 system has proven its worth over time and we now use it so routinely that we do not even think about it.

Playing with Logarithms

Any numeration system—whether base 2 or base 10 or base 15—depends on the use of positional notation. This concept means that

---

*A base 15 system would consist of fifteens, fifteens of fifteens, and so on so that the number 31 in the base 10 system (3 tens and 1 one) would be represented as 21 in a base 15 system (2 fifteens and 1 one).

the place in which a given numeral appears in a number, such as the 3 in 1231, determines the magnitude of that number. In other words, the numeral 3 is in the tens columns and represents three tens. As such, the position of the number has a greater impact on its quantity than its actual face value. The 3 merely refers to some aspect of "threeness" which does not become apparent until the numeral 3 is placed in a specific position in a number. If 3 appears by itself, then it occupies the "ones" position by default and so we know that we are referring to three ones or the amount of 3. If the number 3 is followed by a 0, however, we know that the number in question is 30 and the 3 here refers to three groups of ten.

Of course such a mind-expanding discussion of positional notation can leave even the most brilliant of individuals gasping for breath and literally overwhelmed by the sheer magnificence of the words themselves. Even Shakespeare never attempted to explain positional notation in any of his 37 plays or one of his many sonnets. Perhaps the Bard realized that he could not accomplish everything there was to accomplish in literature and left a dramatic explanation of positional notation to a future successor. But the apparent simplicity of this concept should not confuse anyone as to its indispensability in the foundation of our modern mathematics. Positional notation made it possible to represent large numbers in a convenient form and thus deal with problems involving quantities that were thought to be unimaginably vast by earlier mathematicians.

To whom do we owe the honor of the discovery of logarithmic notation? Most historians of mathematics are inclined to credit the Englishman John Napier whose name is accorded little of the celebrity that surrounded that of his fellow Englishman Sir Isaac Newton even though Napier's influence on the future of mathematics may have been nearly as profound. Of course Newton was no slouch in the sciences and did manage to create what we now know as Newtonian physics which in turn gave rise to a mechanistic cosmology that was to dominate science for more than 250 years. This was admittedly a very impressive accomplishment and not one that could have been attempted by the local grocer or even the head cashier at the movie theater. One must be fairly good with numbers to create an entirely new branch of physics or mathematics. Such accomplishments are also very demanding on one's time as one cannot usually make such an important discovery over a cup of coffee or even during a round of golf. The point of this digression is that Newton was perhaps the only person who ever lived who made such a monumental impact on both mathematics and the physical sciences; Einstein, by comparison, was Newton's equal in physics but did very little in the way of original mathematics.

Napier's name is not so well known by the general public because he, like every other scientist of his day, was destined to be overshadowed by the towering genius of Newton. Napier did have the good fortune to live a generation or so before Newton so he was not troubled by feelings of "Newton envy" and was thus free to engage in his studies of mathematics. A resident of Merchiston, Scotland, Napier published his masterpiece "A Description of the Admirable Table of Logarithms" in 1614. This snappy title did not attract the interest of the masses who, in that day, were more likely to be dropping dead from the various illnesses and diseases that periodically ravaged Europe than rushing into the streets to debate the finer points of specific mathematical theorems. But Napier's tract had an indelible impact on the scientific community because it showed the power of mathematical notation and hinted at the many uses to which mathematics could be put to improve the fortunes of humanity.

Napier himself was born in 1550 into a family of substantial means at Merchiston Castle in what is now Edinburgh. Napier could claim an impressive lineage of ancestors who had occupied positions of prominence in government and industry as well the financial resources which afforded him the leisure time to write about subjects ranging from mathematics to religion. As a member of the landed gentry, Napier was also intimately familiar with the ongoing struggle between the Protestants and the Catholics which would continue throughout his life and certainly shape his own often virulent political writings. As a boy, he was sent to the University of St. Andrews in Scotland, ostensibly to protect him from the violence raging between various religious factions. But young Napier was not so shielded from the violence that he could avoid developing an intense dislike for the Catholic Church, which he viewed as the primary cause of the devastation throughout the land. After completing his education, Napier returned to his home at the age of 21 and composed a popular theological work, *A Plaine Discovery of the Whole Revelation of Saint John*, which lobbed verbal hand grenades at the Catholic Church in general and the Pope in particular, whom Napier considered to be the Antichrist. Although Napier's work was very popular with most government officials and his fellow landowners, it was not widely embraced by the lower classes, who had borne the brunt of the ongoing violence between Catholics and Protestants and desired to see an end to the warfare. But the tract did help to solidify Napier's influence in the Scottish government.

It was not until Napier published his treatise on logarithms known as the *Descriptio* that he obtained the immortality that would have otherwise eluded him had he contented himself with stringing together colorful adjectives to describe the papacy. The logarithm of any number is the exponent in the power of 10 which gives that particular number. The logarithm of 10 is 1 because 10 raised to the first power (1 is the exponent) is equal to 10. Similarly, the logarithm of 100 is 2 because 10 raised to the second power (2 is the exponent) is equal to 100. But Napier showed that this same process could be used to express any whole number—not merely those that could be evenly divided by 10. Thus the logarithm of 4 is 0.6021 because that is the number we would obtain if 10 is raised to the 0.6021 power. In this way, any large number could be expressed in an exponential form.

Napier's discovery of the logarithm also came at a fortuitous time for the emerging sciences because his work made it possible for complex calculations to be done with considerably greater ease. In particular, Napier's work coincided with Johannes Kepler's calculations of the position of the planet Mars which in turn led Kepler to the discovery of the laws of planetary motions. Kepler's work was, of course, of crucial importance to Newton's later success in fashioning a mechanistic model of the universe governed by Newton's three laws of motion and his universal law of gravitation.

The importance of Napier's work was soon recognized as a remarkable achievement. One of the first persons to see the value of Napier's work was Henry Briggs, a professor of geometry at Oxford, who made the long trip to Napier's castle so that he could personally meet the discoverer of the logarithm and suggest ways in which the system of logarithms could be improved. Briggs's most notable suggestion was that Napier use base 10; it was advice which Napier gladly accepted. Upon his return to Oxford, Briggs produced several tables of logarithms which would soon become widely distributed throughout Europe and make it possible for the Continent's mathematicians to do complex mathematical calculations with much greater speed. Briggs's name became inextricably bound with that of Napier and his logarithms because Briggs, more than Napier, was the champion of the logarithms in Europe and it was Briggs's prominence which helped to facilitate the acceptance of Napier's work.

Napier's work in logarithms was his greatest contribution to mathematics. He would offer a few more snippets of mathematical insight in his *Rabdologia* including two rules for obtaining square and cube roots. But Napier would devote his remaining years to managing the family estate and keeping abreast of new discoveries in the sciences and new fashions in kilts. He would also correspond with many of the leading scientists of his day but would have very little influence over the evolution of the sciences apart from that offered by his logarithms.

More About the Logarithm

So what is this wondrous thing known as a logarithm? It is actually so simple and so obvious that even Briggs was reported to have been amazed that no one had ever thought of it before Napier. But such is the benefit of hindsight. Napier was fortunate enough to have a brilliant stroke of insight that enabled him to consider ordinary arithmetic not in terms of the numbers themselves but instead in terms of their exponents. Napier realized that if 2 were raised to the 0th power, it would be equal to 1 but that 2 raised to the 1st power would be 2, and 2 raised to the 2nd power would be 4, and so on. He proposed that all numbers used in arithmetic operations other than addition and subtraction be represented by their logarithms (powers of 10) and that the operations be carried out in terms of the arithmetic operations of these powers. For those mathematicians who are constantly on the go, whether it be jetting to the Riviera for a conference on group theory or speaking before the General Assembly of the United Nations on Godel's Theorem, logarithms provide a handy time-saving device. Although the number 10 is called the base of the logarithm, any other number can be used as a base. But most people would probably prefer to use 10 as a base due to our aforementioned familiarity with the base 10 system. What makes logarithms such a short-cut for mathematicians is that multiplication is reduced to addition and division is reduced to subtraction. Moreover, the raising of a number to a power is reduced to multiplication and taking the root of a number is reduced to division.

This concept can be better understood if we actually stroll over to the proverbial blackboard and show how logarithms can make a difference in your life. Suppose that I have 100 cookies which I have just purchased from the local Wilderness Girls troop and you have 1000 cookies which you just purchased from an out-of-work investment banker trying to start a new career in door-to-door food sales. If we want to know how many cookies we would have if I ate all 100 of my cookies on a binge and you fell for the banker's very convincing sales pitch 99 more times, purchasing an additional 1000 cookies each time, then we could proceed using our standard multiplication operation which would give us the answer of 100,000. Alternatively, we could use Napier's logarithms and add the logarithm of 100 (the number of times you purchased cookies from the banker), which is 2, to the logarithm of 1000 (the number of cookies you purchased each of those 100 times), which is 3. We know from our remedial college mathematics course that $2 + 3 = 5$. The number whose logarithm is 5 in the base 10 system is 100,000. So we see that merely by adding together the exponents for 100 and 1000 we are able to obtain the answer 100,000. This is of course much simpler than finding a piece of paper and a pencil and engaging in a laborious manual calculation because very few people still know how to do manual calculations due to the popularity of electronic calculators. But not having to burden ourselves with developing basic mathematical skills has provided us with vast quantities of leisure time. We can use this new-found free time to devote ourselves to bettering our television geography skills and cruising the World Wide Web in search of electronic soulmates who are waiting to hear from us so long as we provide a valid credit card number.

In short, we can see that the addition of the logarithms of numbers gives us a nifty shortcut so that once we know that the logarithm of our base 10 number is 5, we can conclude that the answer is 100,000. So you can be very proud that you own a massive pile of rotting cookies and take satisfaction in knowing that you have helped to provide a broke banker with a new moneymaking career.

But we have not yet quite beaten this topic to death because we need to see how far-reaching the concept of logarithms can be when considering the basic arithmetic operations. The division of logarithms is the reverse of the multiplication of logarithms. If we have 1,000,000 marbles which we wish to divide into 1000 groups, we could use the methods employed by so many razor-sharp civil servants and find a big room and begin carrying the marbles one by one over to a single spot until we had collected 1000 marbles together. We would then repeat the same process until we had amassed a second group of 1000 marbles. At this point, we would be well on our way to separating out our 1,000,000 (10^6) marbles into 1000 (10^3) equal piles. Once we had completed this task, we could then quickly count each of the piles and then we would know that 1,000,000 marbles divided by 1000 (groups) would be equal to 1000 (the number of marbles in each group). But by doing our arithmetic the "fun and easy logarithmic way," we could save a considerable amount of time by simply subtracting from the exponent for 1,000,000, which is 6, the exponent for 1000, which is 3. So this operation gives us the exponent of 10 raised to the third power (10^3) which is solved in a flash, unlike the labor involved in separating 1,000,000 marbles, one by one, into 1000 equal groups. But it is certainly much easier to subtract one exponent from another to carry out the division of one number by another number than it is to retrieve pencil and paper and physically divide 1,000,000 by 1000. This is particularly helpful in our sophisticated technical society in which people are much too busy chatting with their therapists and psychics to learn how to perform actual mathematical operations.

That Napier's work in logarithms has proven to be of great value to modern mathematicians and scientists is clear. Although we have contented ourselves with the use of a base 10 system, there are other base systems that have been used. Indeed, the ancient Babylonians used a base 60 system. But as there are very few mathematical historians who specialize in Babylonian numerology and even fewer people who actually care about such things, we do not have any clear reason as to why the Babylonians embraced such a system. Based on our knowledge of anthropology and Darwin's theory of evolution, we can probably rule out the possibility that the Babylonians had 60 fingers and 60 toes as a plausible explanation for their numeration system. Perhaps the Babylonians based it on some aspect of the solar year which is almost equal to 360 days or perhaps it was merely an arbitrary designation by a mathematically inclined Babylonian that stuck.

Why would anyone want to use a system other than base 10? After all, base 10 is familiar and convenient and is not so elaborate as to cause our hands to cramp up when we write out even a 6- or 9-digit number. But it is not perfect. Suppose we wanted to choose base 8. To write numbers from 0 to 7, we would use the same numbers that we always use in base 10, which are 0, 1, 2, 3, 4, 5, 6, and 7. Once we get to 8, however, we have to move from the first ("ones") position to the second ("eights") position because the second position represents the base (in this case 8) multiplied by 1. So the number 8 as we know it in our own base 10 system would appear as the number "10" in a base 8 system because we would have an 8 in the "eights" position and a zero in the "ones" position. Similarly the number 9 as we know it in our base 10 system would appear as the number "11" in a base 8 system because we have one quantity of eight in the eights position and one in the ones position. In order of ascending quantity, the positions would proceed in the following manner: ones, eights, sixty-fours, and so on. The comparable positions in our base 10 system are ones, tens, hundreds, and so on. Each succeeding position is determined by increasing the exponent attributed to the previous position by one. The exponents here are 0, 1, and 2, respectively. Similarly, if we wanted to describe the number 78 as we know it in a base 8 system, we would break it up by placing a 1 in the third position (the sixty-fours position) and 1 in the second position (the eights position) and 6 in the first position (the ones position).

So the number 78 would be described in a base 8 system as 116 which is equal to $(1 \times 64) + (1 \times 8) + 6$. We can convert base 10 numbers easily to numbers in any other base system by subtracting the amount represented by a particular position (eights, sixty-fours, etc.) and then repeating that operation with the remainder for each successively smaller position. In the case of the number 78, we knew that we had far more than 8 and even a little more than 64 so we were able to subtract 64 and mark that amount in the sixty-fours position, leaving us with 14. We then shifted to the eights column and subtracted that amount from 14, marking a 1 in the eights column to indicate that that amount had also been removed, leaving a remainder of 6 for the ones column. Had the number been 86, for example, the only difference is that we would have had the number 2 in the eights position instead of 1 because we would be able to subtract 16 and still have 6 left over for the ones position.

But is this digression into base systems truly necessary for an understanding of higher mathematics? Well, this book was designed to be very efficient with only a few spots here and there replete with feeble attempts at humor but it is fair to say that one must understand the nature of base systems to have a grasp of arithmetic, which in turn must be understood to gain a true understanding of the concepts of calculus. Besides, some base systems such as the base 2 system are of inestimable value to our modern society. After all, the base 2 system is the basis for the binary code which is utilized by every modern electronic computer. What makes the base 2 system particularly interesting is that it can be described using only two numbers: 0 and 1. For those who shudder at having to remember all the numbers between 0 and 9, the base 2 system may be just the thing to bring some much desired simplicity to life. Indeed, the 17th-century mathematician Gottfried Wilhelm von Leibniz, who, along with Isaac Newton, discovered the calculus, was deeply inspired by what he regarded as the divine nature of the binary system. Leibniz probably spent too much time hunched over his writing desk and could have done with a few more walks in the fresh morning air. But there were comparatively few diversions in 17th-century Europe even though the occasional outbreaks of plague and warfare could be counted on to liven up even the most tepid ballroom dance or Continental opera. So it fell to the base 2 system to provide Leibniz with the unbridled joy that only a few mathematicians and perhaps twins and bigamists can experience. Although Leibniz urged that the Jesuits utilize the base 2 system in their efforts to convert people to Christianity in the Far East, it is unclear that the base 2 system was as persuasive as outright imperialism in convincing the Asians of the need to turn the other cheek.

But the base 2 system does have an advantage in that it can serve as a model for computers having incredibly vast speeds and storage capacities. This base system works so well because all computers are essentially gigantic collections of switches, which are either "on" or "off." Happily, the base 2 system offers two elements which can be used to represent the on and off positions. One can record a specific number if one has enough switches. To record the number 15 in the base 2 system, for example, one would need several switches, each of which would represent a given position in the base 2 system. So we would have one switch for the ones position, a second for the twos position, a third for the two squared position, the fourth for the two cubed position, and so forth. The number 15 would be described using four such switches, each of which would be switched in the "on" position to indicate a value in a particular column. The number 15 would thus be represented in the base 2 system as 1111, which, from a positional standpoint, would be equal to $8 + 4 + 2 + 1$. Each of the switches here would be on to indicate that a "1" should be placed in each of the four positions. If we wished instead to input the number 14, we would not activate the "ones" switch but would still activate the other three switches so that we would represent the number 14 as 1110. Certainly one's imagination could revel in the fun that converting base 10 numbers to base 2 numbers would offer to even the most dour of individuals. But this system does provide a very efficient way to describe even the largest meaningful numbers if only one bothers to construct a computer with enough switches to represent that number in a base 2 system. So it should be possible to calculate the product of one zillion times five bazillion if only one can figure out how to describe a zillion or a bazillion in base 2 terms.

But the computers also have memories which make it possible to store the results of previous calculations and combine them, when desired, with other calculations. The computer can therefore combine any of a wide variety of calculating procedures including addition, subtraction, multiplication, and division. Moreover, they can perform these calculations at speeds and accuracies and complexities far beyond the capabilities of even the most intelligent humans. This ability to carry out such impressive mathematical operations has prompted many persons to speculate that computers will one day replace human beings. But these writers have failed to notice several crucial distinctions between human beings and computers. First, human beings do not require batteries or an electrical outlet to run; they operate quite well on a diet of food and water. Second, computers are calculating machines; they do not conceptualize or have unique thought processes which lie outside the parameters of their programs. Human beings, on the other hand, have an infinite capacity for creating original items—whether it be priceless artwork or hazardous wastes. Computers are, in this sense, no different than forklifts. They do the heavy work for human beings but operate according to the directions provided by their human mentors. So in this sense the human being is no more likely to be supplanted by a crane than it is by the computer.

Like computers, logarithms are tools which are used to simplify basic mathematical operations by providing us with efficient and concise ways to organize potentially vast amounts of data. The basic features of all arithmetic operations can, like Euclid's geometry, be reduced to a few rules which are susceptible to infinitely varied manipulations and applications. As such, the usefulness and applicability of even the most basic concepts in arithmetic offer a model worthy of emulation. As we shall see, this lesson has been taken to heart by the discoverers of nearly every branch of higher mathematics including the calculus.

CHAPTER 4

Strange Numbers and Stranger Symbols

They say that habit is second nature. Who knows but nature is only first habit.
—BLAISE PASCAL

No doubt the reader is feeling somewhat relieved that the previous chapters have not delved too deeply into the arcane symbols and forbidding nomenclature that is the province of the mathematician. But our job is not to coddle our readers because they have no doubt lost the sales slip for this book by this time and have no choice but to keep this book on their coffee tables or bookshelves. So there is no need to soften the subject matter for the sake of selling a few copies because very few bookstores will take back a used copy without a sales slip—no matter how persistent your pleas for mercy to the sales clerk may be.

This is one of those chapters which, in any other book, would strike fear into the hearts of readers who are unfamiliar with such mathematical curiosities as irrationals and radicals. But hope is here and there is no need to cower under your bed hiding from the big bad square root or the blood-sucking negative number. After all, there is a great big world of mathematical wonders out there that do not fall within the group of whole numbers which we sometimes refer to as counting numbers (1, 2, 3, 4, and so on). Perhaps the most obvious of these candidates is the square root, which was reportedly created by the Pythagoreans when they were considering the deeper ramifications of the Pythagorean theorem—namely that the sum of the squares of the two perpendicular sides of a right triangle is equal to the square of the third side of that triangle. When the Pythagoreans reversed the process, they were forced to consider the troubling idea that not every right triangle's sides could be expressed in precise whole numbers. In other words, there are some right triangles in which the Pythagorean theorem requires only whole numbers (such as 1, 2, 3, and so on) and other triangles in which the theorem itself requires the creation of a different type of number that lies somewhere in between two whole numbers. While such a discovery might have dismayed some persons, the Pythagoreans were overjoyed to see that they would have a very interesting mathematical puzzle to consider, particularly as they were not at that time being persecuted by the local natives and fleeing for their lives. The Pythagoreans discovered that their theorem worked quite well with triangles where the two known sides were, for example, 3 and 4 units because they could add the squares of these two numbers and determine that the third side was equal to the square root of 25, which is 5. But this approach led to more unsettling results when other right triangles were used, such as those having sides of 1 unit on each of the two perpendicular sides. The squared sums of the first two sides would be equal to 2 because $1^2 + 1^2 = 2$. But to search for the square root of 2 created an unanticipated headache because $\sqrt{2}$ is not equal to a whole number. The Pythagoreans did carry out numerous calculations to see if $\sqrt{2}$ was some exact fraction in between the whole numbers 1 and 2 but this hope was never realized. Instead the Pythagoreans gradually accepted the fact that there was no single fraction which, when squared, would be equal to 2. They also devised a detailed proof which was later described by Euclid in his *Elements*. The net result of this very elaborate exercise was the determination by the Pythagoreans that $\sqrt{2}$ cannot be formed from the ratio of two whole numbers. We could, of course, describe this proof in elaborate detail but we realize that our readers are action-oriented and would not necessarily sit still for a tedious mathematical proof. They are also open-minded and willing to take our word for it that $\sqrt{2}$ is not a number that can be formed by dividing any whole number by a second whole number.

One can only imagine the consternation that this discovery caused the Pythagoreans who, quite frankly, were already under quite enough stress trying to hold together a community founded on numerology and free love. Their entire philosophy of mathematics was based on whole numbers and they had, through their own investigations, torn a gaping hole in this philosophy by creating what would become known as an irrational number. Now there are many people who do not believe there is anything rational about higher mathematics but the term irrational has nothing to do with thought processes; it was a term coined by the Pythagoreans to refer to an unknowable number because of their inability to calculate it using whole numbers. According to popular legend, the members of the Pythagorean sect who discovered $\sqrt{2}$ were on a ship at the time and were rewarded for their ingenuity by being tossed overboard to the sharks by some other Pythagoreans. These Pythagoreans apparently did not consider the discovery of $\sqrt{2}$ to be a good thing and, like most sect members, were not particularly tolerant of dissenting ideas.

But it was hard to keep the discovery of irrational numbers under wraps and word soon got out that there was a new type of number in town. As the subject was explored in greater detail, it became clear that there were many different irrational numbers including the square root of any whole number that is not a perfect square or the cube root of any whole number that is not a perfect cube. Indeed, there were more irrational numbers than there were whole numbers, so the can of worms opened by the Pythagoreans turned out to be very large.

Although some of the more reactionary Pythagoreans might have liked to have continued throwing their more inquisitive members overboard, others recognized the futility of trying to preserve the older mathematics, which did not allow for the existence of any numbers other than whole numbers. As a result, the Pythagoreans decided that they needed to broaden their rather narrow concept of numbers to include non-whole numbers such as irrationals. But this new expansion of the Pythagorean numerology was not so straightforward because the Pythagoreans had this bizarre new concept of number but they did not really know what to do with it at first. They hit upon the idea of returning to the operations of addition, subtraction, multiplication, and division with whole numbers and seeing if such operations could be carried out using square roots. But this decision probably was made only after numerous attempts to approximate these square roots with fractions or decimals. Unfortunately, the Pythagorean theorem does not allow for approximations—for a right triangle having two sides of 1 unit of length, the third side must invariably be equal to $\sqrt{2}$. One cannot satisfactorily say that "1 squared plus 1 squared is sort of equal to 1.41." Words like "sort of" and "kind of" will drive many mathematicians into fits of rage and they will not hesitate to curse the ancestry of those who would settle for inexact approximations in their calculations. In short, mathematicians would rather have a stable of accurate, irrational numbers (even though these numbers do not have precise positions on the number line) instead of less exact approximations. Remember that the ultimate integrity of the mathematical process is what is most important to mathematicians; the fact that we cannot easily imagine the number itself or the ways in which it may be used is of secondary importance.

But the irrational number was something of an orphan for many years following its discovery by the Pythagoreans. Students and professional mathematicians alike were openly hostile to it, in large part because they did not understand its role or the reasons underlying its operations. Mathematicians were far more interested in developing new branches of mathematics such as the calculus and devising practical applications for these discoveries such as celestial mechanics. The irrational number remained on the outskirts of mathematics, more often ignored by mathematicians or, at best, treated as something of a curiosity. This is not to say that the irrationals were complete pariahs. Geometers found it easy to use them when dealing with the Pythagorean theorem and right triangles. After all, one could easily have a right triangle with an irrational (square root) for the hypotenuse because of the basic requirements of the Pythagorean theorem. In a right triangle with two sides of 1 unit in length, the third side, as we have seen, would be equal to $\sqrt{2}$. Certainly this is not an earthshaking statement, particularly to those readers who are closet Pythagoreans, who embrace both triangles and the doctrine of free love with unbridled passion. But in this case the irrational was viewed as little more than a label to be placed when needed on the third arm of a right triangle. Viewed purely in terms of a triangular leg, it did not seem to trouble mathematicians very much. But this cursory use was deceptive in that it made it possible for mathematicians to shunt irrationals to the side and avoid thinking about the potentially troubling ramifications of non-whole numbers.

This confusion about what to do with irrational numbers began with the Greeks who followed the Pythagoreans and was merely compounded in the ensuing centuries by equally ignorant mathematicians. These Greeks, for example, found themselves unable to utilize irrationals in any practical way because they viewed them as static expressions much like the triangles that dominated their geometry. Irrationals might appear as symbols describing the length of a hypotenuse but they would have little more relevance to mathematics than a portrait hanging on a wall. Moreover, the Greeks avoided the use of irrationals altogether in their calculations, relying solely on whole numbers and fractions. This attempt to sidestep irrationals was ultimately doomed to failure because it necessarily limited the types of practical problems which the Greeks could solve.

Although the Renaissance marked a flowering in the arts and the sciences in Europe, it did nothing for irrational numbers. Mathematicians continued to avoid irrationals like the bubonic plague which periodically pruned their ranks. But unlike the plague, which caused Europeans to drop like flies, mathematicians still felt free to deny the existence of irrationals. Indeed, one would not be surprised to learn that the great French philosopher René Descartes, despite thinking that he existed, did not extend the same courtesy to irrationals. Descartes's pessimistic view of irrationals was shared by English mathematicians such as Isaac Barrows, the Lucasian professor of mathematics at Cambridge University. Try as most mathematicians might to deny the importance of irrationals, they were nevertheless finding their way into all sorts of nooks and crannies of mathematics, ranging from algebra to calculus. But the use of irrationals tended to be confined to operations which basically mimicked those which utilized whole numbers. It was only as time marched on and these fuddy-duddy reactionary mathematicians died, freeing up the universities for younger scholars with more tolerance for new ideas such as irrationals (which had only been lingering around for 16 centuries or so), that mathematicians began to undertake serious efforts to devise formal rules for handling irrationals in mathematics.

While the Europeans did their best to pretend that irrational numbers were nothing more than a figment of their imaginations, mathematicians on the Indian subcontinent were devising the equally revolutionary and, in some ways, equally inscrutable negative number. Why this innovation came from India as opposed to a Europe which was just beginning its efforts to colonize those parts of the world which it did not actually own is unclear. Perhaps the Eastern emphasis on unity and periodically recurring phenomena prompted Indian mathematicians to look for a negative counterpart to balance positive numbers. Perhaps one particularly insightful Indian mathematician had some extra paper left after drawing a number line and decided to continue extending it to quantities less than zero. It is likely that negative numbers originated with the recognition that they could be helpful in handling everyday mathematical problems. For those individuals who wished to account for both their income and their expenditures, negative numbers proved to be a very useful tool. On those days when a sultan wrote too many checks and bankrupted the kingdom's treasury, the concept of negative numbers might have proven very handy. After all, the official reconciliation of the account might show that the sultan had spent more rupees than he had in his account, thus showing a negative balance. In these olden times, however, the sultan could always have had the official accountant beheaded so that no one would be the wiser as to the country's finances. But negative numbers could certainly be helpful for a day-to-day analysis of income and expenditures, with income such as plundered treasures from neighboring kingdoms being shown as positive entries on the balance sheet and expenditures on elephant diapers being shown as negative entries.

The use of negative numbers also made it possible to deal with the concept of net worth. If the Sultan of Kashmir, for example, had 10 million rupees in his treasury but also owed 1 million rupees to the Sultan of Bombay due to his having made an ill-advised bluff in the previous week's poker game with the other local monarchs, then we can see that both positive and negative numbers can be meaningfully used. As the Sultan of Kashmir has 10 million rupees, this is a statement of his positive assets or wealth. But the debt owed to the Sultan of Bombay may be represented by a negative number equal to 1 million rupees. We can then determine the Sultan of Kashmir's net wealth by subtracting the 1 million rupee debt from the 10 million rupees of assets which gives us a figure of 9 million rupees. Now 9 million rupees is nothing to snicker at, but the Sultan of Kashmir cannot but feel envious of the Sultan of Calcutta, who boasts some 30 million rupees in his national treasury and has a different silk outfit for every day of the year. Indeed, the news of the Sultan of Calcutta's overflowing treasury might so depress the Sultan of Kashmir that he would be tormented by bad dreams at night about being a beggar on a city street wearing a sign marked "Homeless Sultan—Will Work For Food" on his turban.

The concept of negative worth might prove to be the thing which restores the Sultan of Kashmir's faith in the old adage of "what goes around, comes around." For a closer analysis of the ledgers kept by the Sultan of Calcutta might reveal that the Sultan has incurred enormous debts to his tailors and to the dressmakers who ensure that his eight hundred wives are dressed in the finest garments. Even more devastating would be the revelation that the Sultan of Calcutta's new palace had, in the finest tradition of government building projects, incurred enormous cost overruns which were so huge that they could not be disguised regardless of how many royal accountants the Sultan put to death. Imagine the joy the Sultan of Kashmir would feel when he learned that the Sultan of Calcutta actually owed 40 million rupees to his creditors so that his net worth was actually −10 million rupees. This would mean that the Sultan of Calcutta, despite his 150-room palace, his obvious enthusiasm for polygamy, and his colorful silk wardrobe, was worth 10 million rupees less than the beggars who lounged around the front of the palace gates, selling maps to celebrity homes or leprosy ointments for non-lepers "just to be safe." Thus the Sultan of Kashmir would learn first-hand that the concept of net worth is much more meaningful than the concepts of assets or debts separately because it is the only one of the three concepts which provides a summary of the true financial condition of the Sultan of Calcutta. So the Sultan of Kashmir would consider himself to be quite the financial powerhouse even though his 95-room palace was comparatively modest (but only with a single mortgage as compared to the three encumbering the Sultan of Calcutta's home) and he could boast only 38 wives (even though he did know all of their names and the locations of most of their tattoos).

The use of negative numbers was particularly important in the rise of banks because it made possible the lending of vast amounts of money to people who were not able to manage their money very well so that they could go deeper and deeper in debt and thus help to prop up the national economy. Here, the use of negative numbers to represent multiple debts turned out to be of great use because it enabled any nasty moneylender to show in a very concise way the total amount of debt owed by a particular person. This idea can be better illustrated by telling the story of Dr. Dave, a college professor of literature who dabbled in his spare time with various inventions such as a Taoist lawnmower which did not actually cut grass like a traditional lawnmower but instead emitted mystical chants until all the blades of grass in the yard curled up into compact lotus positions. Dr. Dave had also invented several unique foods including breakfast cereal made of recycled bean sprouts and popsicles flavored with old spinach. The fact that these inventions had never really caught on with the public had not discouraged Dr. Dave in the least as he believed that it was only a matter of time before the rest of the world caught up with his visions. But Dr. Dave had incurred significant costs in the design and manufacture of his prototypes. In fact, Dr. Dave had been forced to go to Gigantic Bank on several occasions to borrow money to fund his research. The bank, for its part, did not think very much of Dr. Dave's inventions, but its staff did appreciate Dr. Dave's frequent efforts to increase his debts. In January, Dr. Dave had borrowed $10,000 to build a 38-seat bike that was almost as long as a basketball court. In March, Dr. Dave had borrowed $15,000 to build ten sets of prototype "clown-sized" athletic shoes, believing that he could tap into a market largely overlooked by traditional shoe manufacturers. Finally, Dr. Dave took out his third and final $5000 loan in June to finish perfecting a cologne which smelled of old fish wrapped in newspaper which he called "Fisherman's Wharf," a sort of brawny, in-your-face type of scent which he expected would have an explosive impact on the perfume industry.

Because each of Dr. Dave's debts could be represented by a negative number, it was very easy for the bank officers in charge of reviewing customer account records to summarize the state of Dr. Dave's overall portfolio with the Gigantic Bank. The three loans, leaving aside whatever negligible amounts had been paid since the loans were closed, would be entered on the bank's ledger as follows: −$10,000, −$15,000, and −$5000, which would total −$30,000. Now such a large sum might weigh heavily on the minds of many persons but Dr. Dave is not one to worry about such mundane things, particularly as he believes that the federal income tax is illegal and that all loans are actually gifts that do not have to be repaid. The important point is that negative numbers can be used to show a lender very easily the status of a borrower's portfolio with the bank, no matter how many loans he or she may have outstanding at the time of the review.

This example can also be tied into the concept of net worth because we merely have to take the sum total of Dr. Dave's indebtedness (including any other liabilities such as the amounts owed on his house, his car, and his credit cards) and subtract that sum from his total assets to learn the amount of his net worth. Even though Dr. Dave may have borrowed money from many banks and unsavory individuals, the exercise will remain the same and we will merely have to subtract the amounts of each loan from his wealth to determine his net worth. This process also works in the reverse because if Dr. Dave is lucky enough to sell his "Fisherman's Wharf" cologne to a major cosmetics company for the princely sum of $250,000, then Dr. Dave's net worth will presumably increase by that very same amount. The auditors at the Gigantic Bank would show this change in his net worth (assuming he was gracious enough to pay back all three of the loans) by canceling out the −$10,000, −$15,000, and −$5000 loans. This would, however, reflect no change in his net worth because although the loan debts of $30,000 would be gone, Dr. Dave would have $30,000 less in cash. Of course Dr. Dave might have become somewhat hostile to the Gigantic Bank after having received numerous nasty letters complaining about his failure to make his monthly loan payments and so he might not be terribly concerned about repaying the loans. Instead he might choose a less honorable course of action such as buying a cabana in the Cayman Islands and somehow learning to live with the guilt of having swindled the Gigantic Bank out of $30,000.

The usefulness of negative numbers extends beyond financial considerations. One can use negative numbers on a historical time line if there is a single date such as the birth of Christ which defines the numbering of subsequent years. Someone who refers to the year "1998" is referring to a year that is 1998 years after the birth of Christ. But what can a shoe historian who is interested in tracing the evolution of the sandal back to antiquity do to demarcate events that occurred more than 1998 years ago? The easiest thing would be for the historian to simply add together the 1998 years which had ensued since the birth of Christ to the number of additional years prior to the birth of Christ at which the specific event occurred. This would require the historian to chronicle his discoveries in the following manner: "Some 2900 years ago, the first sandal was invented by a wandering animal skin merchant who, after having burned his bare feet repeatedly on the hot sands of the Arabian desert, used twine to tie two lamb skins to the bottoms of his feet. Two hundred years later, an Egyptian merchant named Ed attached leather straps to the lamb skins so that they would not slip when he tried to outrun invading mounted horsemen who thought it sporting to spear pedestrians. But it was 2450 years ago when a swordsmith named Oscar broke a molten blade on his anvil and accidentally invented the first buckle. After the passage of a few years, Oscar decided to attach the buckle to the leather straps of his sandals and thus ushered in the modern footwear era."

Although this account of the sandal makes for fascinating reading, it does have a certain clumsiness because one is continually forced to do a little mental arithmetic to figure out how long before the birth of Christ each of these events occurred. Fortunately, we do have negative numbers to provide us with a quick way to ascertain this answer. Once we settle upon the birth of Christ as the equivalent of "0" on a number line, then we can easily go further backwards in time and reference any prior year by the number of years it precedes the birth of Christ. So if our sandal historian is compiling his research in the year 1998 and he is writing about the invention of golf shoe sandals with metal spikes which he believes occurred 220 years before the birth of Christ, then he could designate that year as "the year −220" because it is 220 years before our defining event. But historians have instead opted to express such "negative" years by attaching the letters "B.C." after the year to indicate its time of occurrence before the birth of Christ. So the year golf shoe sandals were created would be expressed by most historians as 220 B.C.

Although most people can understand the idea of debts being shown as negative numbers and events occurring before the birth of Christ as being negative years, there is still some resistance to the idea of negative numbers. Part of this difficulty may stem from our natural tendency to associate numbers with countable objects such as fingers. We can easily associate the number "10" with the number of fingers on our hands because we use the number "1" to denote the first finger, the number "2" to denote the second finger, and so on, until we have reached the number 10 and finish counting the number of fingers on our hands. These objects we are counting are tangible and real. When we think of negative numbers, however, we lose this connection because it makes no sense to talk about having "minus four fingers"; negative numbers cannot be utilized here in a meaningful way. Because most people think of numbers in terms of their associations with tangible objects, they have a difficult time considering negative numbers as being "real" numbers. It seems that a negative number should be associated with an extra worldly object like the fingers of a hand in a parallel universe ("antifingers?") but the reality is far less dramatic. Negative numbers are simply another type of number which is distinguished from a positive whole number by its position on a number line. They have no special powers nor do they promise the secret of everlasting youth nor even a trifecta at the local horse track. Yet they do provide us with a way to see situations in a much more multi-faceted way as shown by our analysis of the concept of net worth which went beyond the idea of simply stacking up a monarch's assets or liabilities in isolation without taking the two into account simultaneously.

Symbols and Numbers

Many people believe that mathematicians take an almost perverse delight in fashioning imposing strings of unintelligible symbols to represent abstract, obtuse ideas which are far beyond the grasp of ordinary mortals. While it is true that there are a few mathematicians who worship the Marquis de Sade and take great pleasure in inflicting their field's most intimidating equations on unsuspecting innocents, the vast majority of mathematicians are not regular practitioners of sadism despite the protests of most students in introductory college calculus classes. But while higher mathematics is replete with all sorts of ornate symbols, these symbols do serve an important purpose in helping mathematicians to express certain ideas in a concise and simplified manner. This is not something that is known to most persons just beginning to study mathematics. Indeed, the first response is often to voice complaints about the work involved in learning an entirely new language in order to even be able to understand the class lessons.

There is admittedly an element of arbitrariness to most mathematical symbols which defies any logical explanation. The "+" symbol, for example, does not really evoke the idea of adding numbers together. Similarly, the "×" symbol which is used to describe the operation of multiplication does not compel one unfamiliar with its assigned meaning in mathematics to think of multiplication. As a result, we are forced to admit that most mathematical symbols are purely arbitrary and must be remembered. Any attempt to divine some underlying master plan behind the nomenclature of mathematics is bound to be a waste of time because there is no fundamental symbolic unity.

But there is a point to this arbitrariness: the creation of a non-literary language that can express complex ideas with the minimal amount of language. Algebra lacks the romance of French or the sensuality of Italian or the passion of Spanish because it is essentially a language of quantities. While it is perfectly adequate for expressing quantitative terms such as the obesity of one's boss, it is admittedly a poor choice for trying to win the affections of the fairer sex. After all, one cannot stand beneath the balcony of a window and expect to win the love of a lady as did Romeo by reciting the Pythagorean theorem or the square roots of the first ten prime numbers. Some women might be impressed with such a feat but they would still be looking for a more sentimental expression of love which, for example, compares their eyes to limpid pools of deep blue water or their hair to the softest silk or their feet to very small canoes. They do not want to hear abstract equations, even if they are sung to the music of a stringed instrument or even a snappy, toe-tapping jig.

Surely mathematics must offer a universal language that can express all ideas in a symbolic, no-nonsense form! Perhaps even all of Shakespeare's plays could be written out in the unintelligible symbols that infest higher mathematics. Perhaps. Although one could easily express in mathematical terms the famous monologue in *Hamlet* which deals with the derivation of trigonometric functions or enjoy the fight scene in *Macbeth* which was choreographed so as to reveal the solutions to the binomial theorem when viewed from the mezzanine, mathematical symbolism is simply not up to the task of conveying the range and power and depth of emotions of the Bard's masterworks. It does not allow for the subjectivity or the ambiguity of the human condition. Instead it is cold and unforgiving of those who lack either the discipline or the precision of thought to marshall its power.

Mathematicians wandering through the dusty recesses of algebra use symbols instead of words wherever possible. A mathematician might state that the letter $"a"$ shall represent any odd number. This is a statement of generality which not only provides a definition for a as it appears throughout the mathematician's analysis but also promotes efficiency because the mathematician does not have to continue offering the same definition over and over again every time he refers to that symbol. Imagine what a wonderful world it would be if every politician simply defined the twenty or so major promises he or she planned to break after the election in terms of algebraic symbols. A reference to *Promise A* by an incumbent U.S. senator, for example, would thus spare the television audience from a collection of stock phrases, cliches, and mangled statistics while cutting the average debate down from several hours to the time it takes to heat a bag of popcorn in a microwave oven. Of course the opportunities for standing ovations and tumultuous cheers would be greatly reduced because few people would be so inspired by a reference to *Promise A* or *Promise B* that they would leap to their feet. But such is the price one must pay in trying to put one's political system on a sound mathematical footing.

Nearly every student who has attended a class in higher mathematics is familiar with the sinking feeling that one gets in the pit of one's stomach when perusing the pages of the class textbook for the first time. If it is a particularly difficult textbook, there may literally be pages and pages of equations consisting of very impressive arrangements of symbols and alphanumeric expressions which seem to defy interpretation. The easiest solution is, of course, to drop the class and find a major in the liberal arts which will promise several years of personal education—if not actual employment following graduation. But one should confront the apparent mystery posed by these symbols and, like the neophyte who wishes to speak French or Italian or Spanish, learn the meanings of these symbols and the rules that define the ways in which they are used. In this sense, learning the language of mathematics is not very different from learning to speak a foreign language. Of course you cannot impress a date in a fancy French restaurant by telling the snobby waiter the value of π to 38 decimal places because it serves no tangible culinary purpose. But the mathematical language is more rigid than any spoken language and, in a sense, has fewer words and fewer "grammatical" rules to learn. Most spoken languages have an annoying tendency to mutate over time as words and phrases from other languages or terms which have been coined to designate new inventions and technologies infiltrate them. Mathematics, on the other hand, is less susceptible to such influences because it is much more self-contained and, in many ways, much more a prisoner of its own orthodoxy. So it is certainly arguable that one who masters a second or third language should have an easier time mastering the language of mathematics because it is more compact and ossified.

Did Archimedes wallow in mathematical symbolism? Not really. Like most of his fellow mathematicians from antiquity until well into the Renaissance era, Archimedes relied on numbers and equations. His mathematics was largely devoid of the ornate symbolism that would begin to ensnare the subject in the 16th century. But the fact that mathematics became more complex was not due to a conspiracy by the mathematicians of the day to make their subject even more inaccessible to the general public. Instead it was a by-product of the increasingly complex problems which mathematicians were being asked to solve as the embryonic science of Kepler, Galileo, and Newton gave birth to a bewildering array of new technologies.

One of those persons who is most responsible for clothing mathematics in the fancy garment of symbolism was Francoise Viete, a lawyer by training who hobnobbed with King Henry III during the day and devised algebraic symbols with the glee of a diabetic running loose in a confectionery shop at night. Viete, though schooled in the law, was more interested in immersing himself in dense mathematical treatises and cracking the coded messages sent by France's mortal enemy, Spain, to Madrid's confederates on the Continent. The apparent ease with which Viete was able to unravel the Spanish codes won him admirers both in Paris and Madrid. It was facilitated in part by Viete's ability to see repetitive patterns in the seemingly incongruous arrangements of letters appearing on the Spanish cables. Perhaps it was this ability to superimpose order upon apparent chaos which prompted Viete to become enamored with algebraic notation. In a sense, algebra provided a sort of code whereby generalized expressions could be described with very little effort and in very little time. Given Viete's facility with codes, it was not surprising that he would play an important role in the development of an algebraic alphabet. His principal contribution was to encourage a greater use of single letters to represent entire classes of numbers so that a series of manipulative equations could be carried out using the letters as a sort of shorthand. So Viete could let "b," for example, represent the class of odd numbers and then merely reproduce "b" wherever it was supposed to be in the ensuing set of equations.

Even though this use of alphanumeric notation made it possible for laborious calculations to be carried out in a comparatively brief time, most mathematicians of Viete's time did not embrace the new notation. Their reluctance to use letters to represent entire classes of numbers was due in part to the inherent conservatism of mathematical scholars as well as the satisfaction felt by many that the tried-and-true brain-numbing methods for carrying out computations were perfectly satisfactory. Nevertheless, Viete persisted in what would become a lifelong struggle to get his more pigheaded colleagues to at least consider the advantages of single variables in complex equations. Viete did make some headway but progress was slow. Sadly, he did not have the power to sign death warrants and thus encourage the acceptance of his ideas in the universities throughout Europe. But by the time he drew his last breath in 1603, the idea that symbols could play a role in expressing mathematical relations had obtained a foothold in the dialogues being carried out by mathematicians.

Viete's determination to recast mathematics in concise symbolism was driven by the need to solve problems involving unknown quantities. Here is where the use of variables such as x and y could be used to stand for unknown quantities. An unknown quantity could then be solved by placing all of the unknown values on one side of an equation and all of the known quantities on the other side of the equation. Upon the uttering of a few magic words and the use of a few operations to simplify the equation, one can boil the equation down to its most basic form and, if all went well, solve the unknown values.

Of course this process is a little more involved than a single pithy (but well-crafted) paragraph might suggest. But it is not an insurmountable task. As this is a book that is geared toward the wider audience, we must be careful to start at the beginning by viewing numbers and algebraic symbols as being interchangeable. When we say $7x$, for example, we are saying that we are multiplying 7 by some unknown value. The variable x refers to this unknown value. Whether the value for x is 1, 4, or even 18 is not known. However, we do know that algebra is concerned with numerical manipulations so that $7x$, in the absence of any information to the contrary, does not refer to 7 cows or 7 ribbons or even 7 mathematicians. It merely tells us that we will be multiplying 7 by some number. If x is equal to 3, for example, we know that $7x = 21$. "Gosh!" you say. "Tell me more!" Okay. This form of mathematical manipulation saves the investigator from having to reproduce the entire equation which tells us to multiply 7 by x, where x is the value of any whole number.

Now this is a fairly straightforward point, but it does not really tell us anything except that it is possible to place a letter by a number without precipitating hostilities in the Middle East. So we need to take the next step in our effort to become much smarter in mathematics than our neighbors so that we can properly snub them when we bump into them at the grocery store. This next step involves solving the value of x for an equation.

Why would we want to go to all the trouble to solve the value of x in an algebraic equation? Well, the x value is an unknown quantity that makes it possible to express real-world problems so they can be easily solved. But before we can venture into these waters we need to understand what is meant by solving the x value. So without further ado let us consider the innocent looking equation:

$$8x + 3 = 2x + 7$$

For those of you who have not yet left the room in terror, you are to be congratulated for your internal fortitude. But we need to keep in mind that there is nothing unholy about the x symbol. It is merely a reference to an unknown numerical value. So we can think of the x as the number behind curtain number 1 or door number 2, whose identity will be revealed to us in due time.

To solve for x we have to engage in a little bit of manipulation; a talent which should come easy to those of us who have ever completed a federal income tax return. The first step in this operation is to get all of the x values on one side of the equation and the whole numbers (those which are not joined at the hips with an x) on the other side of the equation. But we must remember the cardinal rule of algebraic manipulation that we must always change the sign of a number from negative to positive or positive to negative when we move it to the opposite side of the equation. Of course we need to use this equation to explain what we are talking about.

In starting with our equation $8x + 3 = 2x + 7$, we need to shove all of the x values to one side of the equation. This is sort of the mathematical equivalent of the proverbial high school dance where all the boys stand on one side of the gym and all the girls wait on the other side. We then do our mathematical mixing by shifting $2x$ to the left side of the equation and 3 to the right side of the equation.

Because we must change the positive numbers to negative and the negative to positive when they cross the equal sign (which is sort of a mathematical demilitarized zone), however, we now find ourselves with the following equation:

$$8x - 2x = 7 - 3$$

This equation is equivalent to the first equation even though it appears to be very different due to our shifting of the numbers. So we now need to carry out the two operations in the above equation which involve subtracting $2x$ from $8x$ and 3 from 7. Once we have completed these operations, we will be left with the following equation:

$$6x = 4$$

Things are obviously beginning to reach a crescendo of excitement because we only need to divide both sides by 6 so that we will finally arrive at a value for x. Once we have completed this task, we get the following solution:

$$x = \frac{4}{6}$$

Gulp! A fraction! Well, sometimes we simply have to face our fears squarely and march forward. Our value for x can be simplified by reducing $\frac{4}{6}$ to its most simplified form, namely, $\frac{2}{3}$. So $x = \frac{2}{3}$.

If there is nothing interesting to watch on television, we might as well verify this equation by plugging the value of $\frac{2}{3}$ in each place where x appeared in our original equation to see if it is true. So $8x + 3 = 2x + 7$. By substituting $\frac{2}{3}$ for x on both sides of the equation, we get the following value: $8(\frac{2}{3}) + 3 = 2(\frac{2}{3}) + 7$. Once we have carried out the grubby mathematics, we get $\frac{16}{3} + 3 = \frac{4}{3} + 7$. This is further simplified by multiplying both whole numbers by $\frac{3}{3}$ so that we can stuff the whole numbers into the fractions and compare our answers. So we get $\frac{16}{3} + 3(\frac{3}{3}) = \frac{4}{3} + 7(\frac{3}{3})$ which is equal to $\frac{16}{3} + \frac{9}{3} = \frac{4}{3} + \frac{21}{3}$ which can be simplified as $\frac{25}{3} = \frac{25}{3}$. The fact that both sides of the equation are equivalent to each other is not only a mild relief as it avoids our having to explain why we cannot competently carry out a basic algebraic operation but it also shows us that our solution for the value of x is correct. Had the two sides of the equation not been equivalent, then we would have either stumbled upon an innovative approach to mathematics which prizes anarchy above rules and order or simply goofed up. Unfortunately, the latter explanation would probably be the correct one as mathematicians are not fond of anarchy or of people who would upset the entire edifice upon which the foundations of mathematics has been constructed. In any event, this simple equation should illustrate that we cannot withdraw in horror simply because we are presented with an expression having an unknown variable represented by a letter; we can, through the deft handling of our equations, find solutions to those unknown variables.

Another entertaining form of algebraic symbolism that causes neophytes to run away in terror is the use of multiple variables in parentheses. Perhaps the most common such equation is the following:

$$(a + b)(a - b)$$

Parentheses are not a sign of the beast mentioned in the Book of Revelations but instead a device for directing the order in which particular mathematical operations are to be carried out. To carry out this operation, we need to remember that each variable in the first set of parentheses will be multiplied by each variable in the second set of parentheses. A more helpful way to think of this might be to imagine the a in the first parentheses (a mischievous fellow) tossing a water balloon at the a in the second set of parentheses followed by a second balloon toss at the b in the second set of parentheses. The b in the first set of parentheses would then toss water balloons at those very same victims in the second set of parentheses. Even though this story might end in a blaze of gunfire with federal agents setting fire to a and b's hideout, it will hopefully show the sequence of steps we need to follow. First, we multiply a by a which gives us a^2. Second, we multiply the outer variables a by $-b$ which gives us $-ab$. Third, we multiply the inner variables b by a which gives us ba. Fourth, we multiply the last variables b by $-b$ which gives us $-b^2$. If we lay out these four steps we end up with $a^2 - ab + ba - b^2$. But the middle two terms cancel out leaving us with $a^2 - b^2$. Again, this is a straightforward mathematical operation which we can solve simply by inserting actual numbers in place of the variables a and b. Of course we must remember that whatever number we assign to either the a or the b variable remains the same throughout the equation or else we will find a crowd of mathematicians with burning torches and pitchforks standing on our doorstep.

The important thing to remember about algebra in particular and mathematics in general is that variables are merely symbols which represent numbers. More people are spooked by the appearance of an a or a $-ab$ merely because they believe they are dealing with something other than numbers. But they are dealing with numbers; the identities of these numbers are merely concealed by the letter variables. Perhaps it would be a bit more interesting to think of the numbers as being naked and the letter variables as clothing. Of course you may prefer the voluptuous figure of a 3 to the gracefulness of a y but the important point is that all algebra is numerical operations clothed in the dress of letters.

We shall continue this theme of divide and conquer in dealing with quadratic equations, which sound very scary but are in fact harmless. The wonderful thing about quadratic equations is that they look very imposing to most people even though they are in fact very simple and straightforward. A quadratic equation is usually found in the following form:

$$x^2 + 7x + 12 = 0$$

Of course we can have different numerical values in a quadratic equation other than 7 or 12. After all, this would be a very useless exercise if it only applied to this one single equation. But we shall use this equation to illustrate the point we are attempting to make about quadratic equations. When presented with such an equation, one is typically asked to factor a quadratic equation, which essentially means that we will do the reverse of the operation we just completed and boil this equation down to two sets of parentheses-clad constituents. We already know from the sequence described above that x^2 will be represented by an x at the front of each of two sets of parentheses: $(x\ )(x\ ) = 0$. That is the easy part.

Being mathematical geniuses, we know that we need to find two numbers which, when multiplied together, will be equal to 12. Now there are all sorts of numbers which can fit this requirement including 12 and 1, 6 and 2, and 3 and 4. There are clearly several choices, all of which seem to be appropriate. So what do we do now? Well, we know we have at least three pairs of numbers which, when multiplied, will be equal to 12. However, the winning pair must also meet one additional test: It has to be a pair of numbers that when added together equal 7 because one of our terms is $7x$. Obviously, this will cut down the field of candidates significantly because there is no way that 12 and 1 can be added together to equal 7, even using the most bizarre version of the "new" math. The same point also applies to the 6 and 2 pair. So the only pair of numbers which meets both tests of being added together to equal 7 and multiplied together to equal 12 is 3 and 4. Not surprisingly, these are the factors of this particular quadratic equation. So the factored equation is $(x + 3)(x + 4) = 0$. But we still have to find the value for x which we can now obtain because we know the whole equation equals 0 so that either set of parentheses must also equal 0. Therefore, $(x + 3) = 0$ or $(x + 4) = 0$. The only values for x which hold true are -3 and -4. So -3 and -4 are the two solutions for this equation.

If you really want to have some fun with the members of the local gardening club at their next meeting, you could interrupt the club president's digression on rose fertilizers and explain the addition and subtraction of radicals. No, we are not talking about revolutionaries and anarchists. Radicals are another of those wonderful mathematical concepts which intimidate even the strongest of men even though they are less ferocious than quadratic equations. The important point to remember is that radicals are shorthand expressions for numbers which we cannot easily locate on a standard number line. But there are a few helpful guidelines to handling radicals which do not require us to storm administration buildings and arrest the ringleaders. With algebraic radicals, we have to resist the basic temptation to bludgeon the little tykes into submission and instead adopt a live-and-let-live attitude. Radicals are different from other numbers because they have radical signs in them. So the numbers $\sqrt{8}$ or $2\sqrt{8}$ are radicals but they should be viewed as simply another type of number having a funny wardrobe. But radicals can be handled by any person who is cool and composed. Like radicals such as $2\sqrt{6}$ and $5\sqrt{6}$, which have the same number under the radical sign, they can be added together simply by adding the numerical coefficients in front of the radical sign. So $2\sqrt{6} + 5\sqrt{6} = 7\sqrt{6}$. The same holds true when dealing with subtraction so that $5\sqrt{6} - 2\sqrt{6} = 3\sqrt{6}$. But this plan does not work when we are dealing with dissimilar radicals such as $5\sqrt{6}$ and $4\sqrt{5}$. It is almost akin to trying to subtract apples from oranges or cheese from liver. But some radicals which appear at first glance to be dissimilar can actually be simplified and then recast in similar terms so that they can be added or subtracted. This is a particularly interesting tidbit of information which can be offered to a cashier when one is idling the car at the drive-through window of the local fast food restaurant.

Probability Theory

Once one becomes comfortable with equations and formulas and realizes that they are nothing more than shorthand expressions for straightforward mathematical relationships, then one can grapple with any type of formula without becoming queasy. One of the most helpful formulas, particularly for those who dream of winning great fame and fortune in the casinos of Las Vegas or Monaco, is the probability formula which will give you the mathematical probability of a specific event occurring by merely dividing the desired number of possible outcomes by the total number of possible outcomes in a given event. Although this statement sounds so very complicated, it is actually quite straightforward and can be easily understood if we take the time to consider a few examples.

Suppose that you recently arrived in Las Vegas, having sold your house and brought the proceeds along with you, being absolutely convinced that you are destined to hit it big at the card tables. You might have even purchased a pinstriped suit along with a white fedora hat and matching tie and carnation along with two-toned saddle shoes. Being a professional gambler, you would be cool and calm, concealing your excitement as you stepped up to the blackjack table to try your hand at fame and fortune. You know that you are not a patsy who will soon be separated from his money because you have gone to the trouble to learn the basic theory of probability. You know that you can try to remember the cards that are played in each hand and thereby have a better idea as to the remaining cards in the deck so as to determine when to ask the dealer to give you an additional card and when to hold firm.

After the first hand has been dealt in a game where you are playing alone against the house, you see the dealer revealing a four of hearts and an overturned card. You realize that the dealer has, at best, an ace and, hence, a total of 15. You, on the other hand, have 20 due to your having been dealt two tens. You decide to hold firm because of the slight probability that the dealer will draw a card that will give him the 20 or 21 points the house needs to win the game. If the dealer takes another card (a jack of hearts) and then flips his hidden card over to reveal a king of clubs, then you know that the dealer has a total of 24 points and has thus lost the game. You also make a mental note of the fact that the dealer has now used two face cards. As the game goes on, you continue tracking the number of cards that come up in each hand because you know it will enable you to calculate the probability that a particular card will be drawn in a given situation by the dealer and thus you can make an educated guess as to whether you should hold or take an additional card. How might this information be helpful? Suppose that you have played several hands with the dealer and all of the kings, queens, and aces have been dealt as well as two of the jacks. There are 24 remaining cards and you receive a 10 on your concealed card and an eight on your revealed card. You see that the dealer has received a nine on his revealed card. You know that if only 24 cards are left and there are only two jacks and three tens, for example, then there is only a probability of 5/24 or roughly 20 percent that the dealer can actually beat you with the hand he currently holds in his hand. Knowing this fact, you decide to hold firm and, indeed, throw the rest of your chips into the pot. You congratulate yourself for your firmness even as the dealer reveals the concealed card to be a 10. You stifle the tears and manage to steady the quiver in your voice as you feel very light-headed and fall face-first toward the floor.

Obviously, the previous example illustrates that the calculation of probabilities is merely a calculation—one cannot predict with total certainty that a particular event will occur unless it is the only possible event that can occur, such as the drawing of the ace of spades from a card deck in which all the other cards have already been drawn. But it can provide us with a general idea as to whether a particular outcome is more or less likely than another outcome. As such, it allows us to venture into the den of lions with a cigarette lighter so that we are not totally in the dark about the dangers we are facing. Unfortunately, it will not necessarily prevent us from being eaten.

What are some of the ways this basic probability theorem can be useful to us if we do not have any interest in gambling? Well, it can be useful if one must pick one's undergarments out each morning in a dark room. Suppose that you are feeling in a very festive mood one morning before going to work and you decide to accent your dress with a pair of your fire-engine red underwear. Now you know that your drawer has 18 pairs of underwear, six of which are white, four of which are red, four of which are blue, and four of which are pink (due to the ill-advised washing together of two previously white and two previously red pairs). Because you neglected to pay your electric bill that month to protest the monopolistic pricing policies of the local public utility as well as your having had a run of bad luck at the local casino, you do not have any light to illuminate your underwear drawer. This situation of course requires you to resort to your knowledge of probability theory because you cannot call up anyone else to borrow a flashlight due to the telephone also having been cut off. Anyway, you know that there are four pairs of red underwear out of a total of 18 pairs because you just finished the wash and were forced to match all of the pairs together (even though you cannot figure out where the extra pair of fishnet stockings came from). So you can use the probability formula and divide four (the number of desired choices—the red pairs of underwear) by 18 (the total number of choices—all the pairs of underwear). This division which you were presumably able to do in your head, because you could not find a pencil and paper in the darkness, would reveal that you had a probability of about 22 percent of pulling a pair of red underwear out of your drawer.

But what would you do if you suddenly remembered that you no longer had any blue pairs of underwear due to your having passed them out as "candy" to some Halloween "trick-or-treaters" due to your having run out of goodies early in the evening? Of course this would appear to complicate our probability solution because we must now factor in the loss of four pairs of underwear. But this is easily dealt with because we merely reduce the total number of pairs by 4 from 18 to 14. This means that we have an improved probability of 28 percent of retrieving a pair of fire-engine red underwear on our first try. Of course this probability could be improved much more if the electric bill were to be paid so that your morning adventures in getting dressed would not require you to grapple with probability theory.

Now this is certainly a helpful thing to know but you are probably wondering about how one would calculate the probabilities of multiple events such as the probability of selecting two like-colored pairs of underwear. Even if you could care less about such things, it is important to be aware of the concept because it will help you to be an educated person who can more easily feel superior to others. Let us return to your underwear drawer so that we can get a handle on this concept. Suppose that you have four pairs each of red, blue, white, and green underwear. You take great pride in your collection of underwear and keep them neatly folded in your drawer so that anyone who wishes to inspect your underwear will be very impressed. With a total of 16 pairs of underwear to choose from, the probability that you will pull a blue pair, for example, on your first try while groping in the dark is ¼ (⁴⁄₁₆). Suppose that you decide you want to get a second pair of blue underwear because you will be taking a bus tour and want to take extra precautions in case there is an accident. You know that the probability of pulling the first blue pair is ¼. But then you might think that the probability of pulling the second blue pair is also ¼ because we do have four pairs each of the four colors of underwear. However, this conclusion would be wrong because we will have only 15 pairs of underwear to choose from once we have removed the first pair of blue underwear. Because we know that the probability of pulling one of the three remaining blue pairs from the drawer of 15 pairs is ³⁄₁₅ or, when simplified, ⅕, we can calculate the probability of pulling two pairs of blue underwear from the drawer by multiplying the two probabilities: ¼ × ⅕ = ¹⁄₂₀. So we can see that we have only a 5 percent chance of realizing our ambition of double-layering our underwear in blue today. The important thing we must keep in mind when engaged in such problems is to keep track of the size of the total pool of underwear when calculating the probability of a particular event.

Yet you might wonder how we would calculate the probability of pulling out two pairs of blue underwear if we put the first pair of blue underwear back into the drawer before drawing the second. In this case, we would not be reducing the total number of pairs of underwear available for selection. As a result, we would have four pairs of blue underwear out of a field of 16 pairs for each of our two selections. This means that the odds would not change in moving from the first choice to the second so that we would merely have to multiply ¼ (the probability of drawing one of four blue pairs from a field of 16) by ¼ (the probability of drawing one of four blue pairs from a field of 16 a second time) which gives us a probability of ¹⁄₁₆. We can see that we have a slightly better chance of retrieving two pairs of blue underwear in consecutive draws when we replace the first pair than when we keep the first blue pair out.

The rules of probability are not merely limited to blue pairs of underwear or even clothing but apply to all types of things and can be mastered fairly easily. Indeed, mathematics itself is not completely unintelligible but instead offers its own logical structure and rules of operation which make it at least somewhat accessible to anyone who is willing to devote some time and energy to the task. Certainly it is populated by a gallery of seemingly strange characters and symbols but they are not so fierce and nasty as they might first appear. Indeed, they make it possible for us to see that there are other dimensions to our mathematical universe as well as a nomenclature that is surprisingly elegant, concise, and easy to use. The next chapter will explore a number of applications in which the language of mathematics is used to express complex relationships which are of inestimable value to our daily lives.

CHAPTER 5

More Nifty Formulas and Equations

Work consists of whatever a body is obliged to do, and play
consists of whatever a body is not obliged to do.
—MARK TWAIN

Having stuck our toes into the cold, unforgiving waters of mathematic symbolism, especially as it rears its ugly head in elementary algebra, we shall continue to press forward to examine a number of other applications in which letter variables are used to provide us with information about practical problems. Before we go any further, however, we should point out that nearly every physical or social science is anchored in mathematical formalism because nearly all branches of knowledge rely on mathematical models or expressions to varying degrees. While it is true that no one has managed to describe Shakespeare's *Othello* or *King Lear* using a quadratic equation, there are few areas of knowledge—whether it be economics or sociology or biology or astronomy—that have not incorporated mathematical models or equations into the fold to provide concise representations of the relationships between various phenomena. So while we will never hear Julius Caesar explain the rules for multiplying exponents as he slumps to the ground with a knife in his back, it is surprising to find how prevalent mathematical expressions are in most fields of knowledge.

This chapter will continue to deal with some of the forms in which mathematical equations appear, particularly as they relate to our everyday lives. Obviously, this is the type of hands-on approach that will prompt even the most cynical of readers to leap to their feet in joyous celebration. So perhaps the first place to begin is with those wonderful questions that involve sending different people in different directions at different speeds for differing amounts of time and then asking what they had for breakfast or whether one can determine their religious affiliation. Fortunately, these so-called distance–rate–time problems can usually be disposed of quite easily because one need only be aware of the following algebraic relationship:

$$d = r \times t$$

Those readers who have not been using this book to press their wrinkled dollar bills will quickly surmise that the variables of this equation, d, r, and t, represent *distance*, *rate*, and *time*, respectively. One is often forced to grapple with such problems when taking college entrance examinations and filling out credit applications for snooty department stores. Fortunately, the formula we described above coupled with a clearheaded analysis will make it possible for anyone to answer these types of questions. The ease of use of this formula can be demonstrated if we consider Fred and Marcy, who, despite having ended a brief romance some years ago, have managed to retain a pathological hatred for each other which has made them very competitive in every area of human endeavor. Today they are involved in a friendly road race which provides a perfect setting for the use of our distance formula. Suppose Fred drives an average of 50 miles per hour for 5 hours but Marcy is able to drive 75 miles per hour for 3 hours before stopping to go into a department store for the spring white sale. Who has driven the greater distance? Our formula tells us that the value of d for Fred is equal to 50 (miles per hour) × 5 hours or 250 miles. Similarly, the value of d for Marcy is equal to 75 (miles per hour) × 3 hours or 225 miles. So even though Marcy drove 50 percent faster than Fred, her impromptu stop at the mall occurred too soon in her journey for her to have amassed as many travel miles as Fred. However, she did manage to purchase a pair of very sharp wire cutters which can be used to cut the brake lines of any automobile and a 5-lb. bag of sugar which can be emptied into the gas tank of any automobile requiring an energy boost. Both of these items should come in handy in any future races against

Fred. So Fred's racing effort may ultimately have very little importance in the scheme of things and he may need to start sleeping in his
car to prevent a sneak attack by Marcy.

But this problem is very simple because it involves nothing more
than a straightforward multiplication of two numerical values. What
if we did not have so much information about the race between Fred
and Marcy? If Fred drives 50 miles per hour for 3 hours, we know,
based upon our distance formula as well as common sense, that he
will cover 150 miles in that time. Assume that we are told only that
Marcy travels the same distance as Fred but drives three times as fast
because her Type A personality manifests in a particularly violent
way when she gets behind the wheel of a vehicle. How long will it
take for Marcy to drive this distance? Well, we know that the distance
formula can be solved by substituting 150 for d and 1 hour for t. So
d (150 miles) = r (?) × t (1 hour). The only number which makes this
expression true is a value of 150 for r which means that Marcy
decided to drive at a speed nearly three times that of the posted
speed limit. Fortunately, she did not get a ticket because she outran
the highway patrol and had remembered to wear very tasteful sheer
lingerie under a coat with a matching bullwhip and spiked boot
ensemble to help her talk her way out of a speeding ticket in the
event she was pulled over.

Although this type of equation is very simple, it is an algebraic
expression and, as a result, is no different in concept than the most
complex, convoluted equation that the most twisted mathematical
minds can formulate. It expresses a relationship between three different variables, any one of which can be solved so long as we are able to
determine the values of the other two variables. This is a very fortunate thing for us because mathematics would be of very little use to
us if we were unable to use it to represent various quantifiable
relationships among what may sometimes appear to be totally unre-
lated variables.

There are many practical applications for this type of algebraic
manipulation. Suppose that you are a very wealthy socialite who is
throwing the biggest charity ball of the year. You plan to invite the
bluest of the bluebloods (except for a select few whom you wish to
snub in the most public way possible) and must deal with a staggering number of questions relating to seating, servers, and decorations.

You must also determine how best to handle the logistics of passing out food and beverages. Because you have planned an evening jam-packed with entertainment ranging from an acrobat who juggles dirty laundry to a sword swallower who is deathly afraid of sharp objects, you are very concerned about the amount of time you will need to serve the various courses. You must also make a difficult choice between two very different caterers. The first is the Megahuge Cooking Factory which has a motto promising the "finest in home-cooking" on each of the ten 1000-gallon vats in which it mixes mashed potatoes for its customers every day. Not surprisingly, Megahuge boasts a massive factory in which a bewildering arrange-ment of conveyor belts carries everything from liver paté and rack of lamb doused with mint jelly to truffles and soufflés from the kitchens, which are open 24 hours a day, to a fleet of trucks and transports waiting at the facility's loading docks. You are fully confi-dent that Megahuge can handle the demands of your party but you are concerned that the elegant and exclusive atmosphere which you so dearly wish to maintain will be undermined by its servers de-scending upon the tables in platoon-like formations and unveiling the perfectly square meat and fish entrees which are covered by unerringly equal doses of sauces and which are placed next to identi-cal medleys of vegetables and fruit garnishes, all of which are wrapped in sheets of 100-lb. green cellophane wrap to preserve freshness.

Your other option is to hire Pierre du Mort, who is widely acknowledged by those who pay attention to such things to be the finest chef in the world. However, he is also the snobbiest chef in the world and is so stuck up that he even cooks his own breakfast with a sneer. Chef Pierre also cooks in very small quantities (espe-cially when compared with the likes of Megahuge) and will only prepare one serving at a time. The results of this personal attention are nothing short of a gastronomic orgasm although it may be several days before all of the guests at a major event can be served. But you realize it would be quite a coup to have Chef Pierre ladling some wondrous soup (one bowl at a time) or arranging venison (one chop at a time) or stacking vegetables (one piece at a time) at your party. You would certainly be the toast of the town for many months to come, at least among those persons who were lucky enough to actually get something to eat before the function ended.

However, you are concerned that everyone be served within a reasonable time, and you manage, following the promise of a cash bonus, to convince Chef Pierre to bring along several assistants so as to speed up the cooking and serving of the dishes. But you believe that Chef Pierre will still take longer than Megahuge to prepare the meals to serve your 200 guests. Chef Pierre estimates that he can prepare all 200 plates in 4 hours even though he worries amidst rants of colorful French profanity that the quality of his work will be compromised by such a frantic pace. The officials at Megahuge estimate, by contrast, that they can prepare all 200 plates in 1 hour. The difference between these two rates of production is 3 hours: 4 hours − 1 hour. In its simplest form, we can express the different rates between the times needed for Chef Pierre and Megahuge to prepare the meals as follows: $4x = 200$ and $y = 200$ where x is the number of dishes Chef Pierre can prepare in one hour and y is the number of dishes Megahuge can prepare in one hour. Because Chef Pierre will not budge any further on the issue of hiring additional staff, you reluctantly conclude that you must either have a 4-hour dinner serving or put on an event with a decidedly institutional flavor if only to get through the dinner in a reasonable time so that the evening entertainment can begin with the naked trapeze artists. But then you hit upon an idea after realizing that you are not really too concerned with impressing or even currying the favors of all 200 guests. Instead you decide that there are only about 40 guests whom you really wish to dazzle and the rest, as far as you are concerned, can, in the words of that noted social activist Marie Antoinette, eat cake. So you decide that you will use both Chef Pierre and Megahuge and direct that the plates prepared by Chef Pierre be served to your special 40 guests, with Megahuge passing out the remaining dishes to your "B" list guests. You then wonder how quickly the dinner can be served with the artistry of Chef Pierre and the mass production techniques of Megahuge working in tandem.

Finding a spare piece of paper and a pen, you write out the two algebraic equations $4x = 200$ and $y = 200$. We know that our snippy Chef Pierre can produce 50 plates per hour. We would divide both sides of the $4x = 200$ equation by 50 which gives us a per hour productivity rate for Chef Pierre of 50, so we know $x = 50$. Because we also know that $y = 200$ (which represents the per hour productivity of Megahuge) and that $4x = 200$, we can construct an equivalent equation $4x = y = 200$. This equation tells us that Megahuge's personnel and Chef Pierre's entourage working alongside each other (and trading insults and derogatory comments in two languages) would be able to produce 250 plates per hour. We would express this equation as $x + y = 250$ where x is Chef Pierre's hourly productivity and y is Megahuge's hourly productivity. Because $y = 4x$, we can convert the former expression into the latter so that $4x$ represents Megahuge's hourly production of plates (where $x = 50$) and Chef Pierre's hourly production of plates is 50 (which we represent as x). As a result, we can express this dynamic duo's hourly production of plates as $5x$. As we only need 200 plates, we should be able to serve our 200 guests in less than an hour.

The wonderful thing about algebra is that it makes it possible to calculate with impressive accuracy the exact amount of time it will require to complete the serving of dinner to our 200 guests. But how much quicker will this serving be? Recall that we can serve 250 plates per hour. If $5x = 60$ minutes $= 250$ plates, then $x = 12$ minutes $= 50$ plates. We know that our dynamic duo will complete 50 servings approximately every 12 minutes so long as Chef Pierre does not disrupt the proceedings by launching into a tirade about the homogenization of modern cooking techniques. If we multiply x by 4, which is equivalent to multiplying 50 by 4, this will give us the 200 plates that we need to serve our guests. The expression $4x$ tells us that our 200 plates should be passed out in about 48 minutes. This would be a very happy outcome because we would then be able to bring on the evening entertainment, concluding with the very classy cockfights and the throwing of a beggar to the lions.

This example not only illustrates the author's flawless prose but also the way in which two unknown variables—in this case, x and y—can be handled in the same equation when we can express one variable in terms of the other. When we were dealing with Chef Pierre and Megahuge, we saw that we could express y as $4x$. Once we had an actual number that we could place against either variable, then we could actually define the values of x in terms of y and both in numerical terms. Indeed, we can have any number of different alphabetic variables in an algebraic expression and solve for the unknown values so long as we can express them in terms of each other and, ultimately, in terms of actual quantifiable terms (numbers). So you can horrify your dinner guests by taking a few moments between the evening entertainers to ask them to solve an algebraic equation having two, three, or even more unknown variables.

If you like to put your ear to the ground at railroad crossings to listen for the sounds of oncoming trains, then you might be interested in using algebra to calculate the distance of an object based on the amount of time it takes the sounds to reach your ears. Suppose that you are traveling in your car with a sound engineer who informs you that sound travels through the air at approximately 1100 feet per second. You have always heard about tribal peoples who put their ears to the ground to hear the sounds of approaching beasts. Your engineer friend then informs you that sound travels approximately ten times faster through the ground than it does through the air, or approximately 11,000 feet per second. You are naturally impressed with this difference in velocities and you decide to test this theory when you pull up to the railroad crossing. Armed with this information, you put the car in park and swagger up to the clanging crossing with the red lights flashing their warning. You bend down and put your ear to the rail and find that you can very clearly hear the low rumble of the oncoming train. Were this a scene in a horror movie, you would get your ear or your hair caught in the railway tie or spike and would be decapitated by the oncoming train. But this is a sophisticated experiment and you are quite careful about looking both ways down the tracks before putting your ear to the rail. You listen intently as the sound grows louder and is then compounded by the blast of the train whistle. You then stand upright and watch the second hand of your watch until you hear that very same whistle coming through the air. Your watch shows that 10 seconds elapsed from the time you first heard the whistle through the rail to the time you first heard it passing through the air. Now you have all the information you need to determine the distance of this train. You merely need to be able to incorporate the differing rates at which sound travels through the air and through the ground and the difference in the amount of time it takes the airborne sound waves to reach your ears following the ground-borne sound waves.

Okay, what is the best way to go about expressing this supposedly simple problem? We could try to set up this equation in a straightforward format and solve for the unknown variable, which in this case is the actual distance between us and the train when we first heard its rumble through the rails. As we noted above, the speed of sound through air is about 1100 feet per second and the speed of sound through the ground or, in this case, the rails, is 11,000 feet per second. We first need to describe the distance x in terms of the different speeds of sound. So we would place x in the numerator of two fractions, one with the denominator 1100 and the other with the denominator 11,000 and then subtract the latter from the former with the result being equal to 10 (the number of seconds between the time we first heard the sound of the train in the rails and the sound of the train through the air):

$$x/1{,}100 - x/11{,}000 = 10$$

This equation states that the time needed for the sound to travel through the air minus the time needed for the sound to travel through the ground or, in this case, the rails, is equal to 10 seconds. The algebraic equation thus expresses this rather long sentence in a very economical and exact manner and provides us with a starting point for ultimately solving the value for x.

We first need to multiply the initial fraction by 10 so that we will have the common denominator of 11,000 for both fractions: $10x/11{,}000 - x/11{,}000 = 10$. If we subtract the second fraction from the first, we obtain $9x/11{,}000 = 10$. Now the trick is to get rid of the fraction altogether by multiplying both sides of the equation by 11,000, which yields the equation $9x = 110{,}000$. If we then divide both sides by 9, we will finally obtain a value for x which is equal to 12,222.22 feet. Fortunately, we have not yet forgotten the point of this exercise which was to calculate the distance of the train when we first heard the noise through the rail. So our result tells us that the train was more than 2 miles away when we first heard the sounds because there are less than 12,222.22 feet in 2 miles (where each mile is 5280 feet).

We should be clear that this method can be used for all types of vehicles including trucks, automobiles, motorcycles, and buses. But one must remember that it is very risky to drop to one's knees in a busy city street and place one's ear to the ground in the hope of detecting an oncoming vehicle. Needless to say, the practitioner of such investigative methods may find his career cut short by the wheel of a truck mashing his head into the pavement. But this book does value the idea of educating the public as to some of the dangers lurking in the shadows of the outside world and so we must point out even the more obvious perils when engaging in such behavior. The point is that algebra makes it possible to take a few pieces of information and then draw out a great deal of information based on the various manipulations we can make in the equations themselves.

Algebra has many practical applications other than ensuring a steady supply of jobs for mathematics teachers. More than one student believes that it is an insidious tool which can be used to reduce even the bravest of persons to a mindless mass of quivering jelly. But it can be used for many purposes other than merely terrorizing unwary students. For example, a few basic algebraic equations can be a very powerful tool in determining how best to use a given amount of building materials for the most efficient design possible.

Suppose that I recently had the good fortune to buy a pile of bricks and a bucket of mortar from a door-to-door salesman and I decided it was high time that I construct a greenhouse. Because I have no glass and have run out of money, this will be a very special greenhouse with concrete block walls and a cement ceiling. Even though it may be worthless as a traditional greenhouse because it will be totally dark inside, I may still be able to use it for growing mushrooms. But what I must do before commencing construction is to figure out the appropriate dimensions for the greenhouse. If I am the typical gardener, I want to maximize the ground area that will be contained within the walls of the structure. If I am able to determine what size the structure should be, then I can better determine how best to allocate my limited quantities of building materials. I do not want to build a very long wall with the enthusiasm that comes with ignorant bliss only to run out of bricks before I have turned the first corner. After all, one cannot have much of an enclosure with only one side. My only hope for growing mushrooms in such a structure would be to plant them in the shadow of the wall. But we could only count on a perpetual shadow if we were fortunate enough to live in a world in which the sun had been blotted out by the dust kicked up by a meteorite hitting the earth. So we would not want to make the mistake of constructing a single wall unless we decide to abandon the concrete windowless greenhouse plan and opt instead to build our own version of the Great Wall of China.

But let us assume that I still want to build a concrete structure because I have very little going on in my life at the time and need a hobby. I am naturally attracted by the prospect of using mathematics to utilize my building materials in the most efficient manner possible, so my next step is to retrieve a handy textbook and try to calculate the optimal dimensions for my concrete greenhouse. If I have enough bricks to build a single wall which is 400 feet in length, then I know that the largest possible perimeter of my structure will be a total of 400 feet. This means that the sum of the lengths of the structure will be equal to 400 feet. My razor-sharp intellect also tells me that any two adjacent sides of this structure will be equal to 200 feet in length. In other words, the sum of the length and the width of the building will be equal to 200 feet. Because we are able to call the most brilliant mathematicians in the world, we can determine the length and width of our building without resorting to the less than satisfactory method of making wild guesses as to the appropriate dimensions. Although wild guesses may appeal to television audiences watching game show contestants locked in mortal combat over the right to claim a ceramic pig stuffed with bacon, they do not do a lot of good for the reputation of builders who must rely on something a bit more substantial than a hunch when constructing such things as suspension bridges and skyscrapers.

In calculating the optimal sides of this building, we need to start with the little information that we do have. We know, for example, that 200 feet is the sum total of the length and the width of the largest possible building we can build. Let the width be x. We can then represent the length of any single side by the expression $200 - x$. In addition, we know that $x(200 - x)$ is equal to the area to be enclosed. As we have always been told by architects and other people who draw shapes for a living that the square is the most efficient use of space, we could make the plausible assumption that each of the four walls of our building will be about the same length. We can test this notion fairly easily by taking a few sample pairs of numbers which add up to 200 and multiplying them by each other to see which pair yields the largest square footage. If we begin with the assumption that the square is the most likely outcome, we know that each side will be equal to 100 feet because we have already been told that the sum of two sides is equal to 200 feet and the perimeter of the area is 400 feet. Our intellects reveal to us that the only shape that can fulfill these requirements is a square that measures 100 feet in length on each side. We also know that we can obtain the area which would be enclosed by this structure by multiplying the lengths of two sides (in this case, 100 feet by 100 feet), which gives us an enclosed area of 10,000 feet.

Good gracious, what brilliance! One can almost feel the wisdom of the ancient Pythagoreans emanating from these pages. But we can pat ourselves on the back knowing that we have mastered basic multiplication after having trudged nearly halfway through this book. But we need to test several other sample pairs to see if our hunch about the efficiency of the square is correct. So having tried 100 and 100 we can then try 99 and 101. If we add 99 and 101 we see that it yields a total of 200—just like the 100,100 pair. But we find that if we multiply 99 by 101, we get an area of 9999. We know from our many remedial college mathematics classes that 9999 is less than 10,000 by exactly 1!

What if we take a different approach and pick a pair of numbers such as 198 and 2? This would certainly create a very narrow building which would probably not hold the immense hallucinogenic mushroom caps that we would like to grow. Yet, that 198-foot wall would certainly be a very long one and might make it possible to enclose a greater area than that which could be contained in a building which is 100 feet on each side. If we multiply 198 by 2, we find that such a building would be 396 square feet in area. Because 10,000 is greater than 396, we see that the geometrical shape of the square is still the preferable one if one wants to maximize the square footage in this new building. In moving from the 198,2 pair to the 99,101 pair, we see an interesting trend beginning to appear: The greater the difference between the two numbers (the length and width of the proposed building), the less the square footage area obtained by multiplying those two numbers. As the difference in the two lengths approaches 0 and the length and width become equal, the amount of the calculated square footage continues to increase until it reaches a maximum value when the two sides are equal. This startling conclusion is supported by our example in which we demonstrated in rather stark terms that the more closely the length and width of a rectangle approximate each other's dimensions, the greater the area of square footage which can be contained within the segments of that rectangle. In terms of practical, hardheaded guidance, this revelation tells us that we should avoid building very long, skinny buildings because there is very little demand for such buildings unless you want to store very long, skinny objects. This aversion to long, skinny buildings is one reason why you see so many square buildings. Even though these square buildings are much less architecturally stimulating than those long, skinny buildings, they are desirable in that they can be used to store those objects which do not fit inside long, skinny buildings.

Those who know all the ins and outs of algebra would take a more sophisticated approach instead of merely tossing out pairs of numbers until the optimal solution was found. The people who criticize mathematicians—particularly those mathematicians who like to scrawl elaborate strings of alphabetic equations on blackboards—would have us think that algebra has no practical significance. They instead consider it to be little more than tricks with equations and manipulations of variables. But the proper equation can be a wonderful time-saver. Such an equation can describe a relationship which would otherwise be unascertainable if one were content to reach into a hat and pull out random pairs of numbers.

Is there a point to this digression? Certainly. Every word in this book is pregnant with meaning and dripping with masterful insights into the human condition. This point about the power of a single equation to describe and predict mathematical relationships as opposed to the hundreds of hours of mindless fun that could be had by random guessing can be further underscored by returning to our example involving perimeters and shapes.

Our basic hunch about the square being the most efficient geometric shape for purposes of maximizing square footage can be considered by using the following equation: $10,000 - x(200 - x) > 0$. This equation offers an expression of inequality which, if true, would prove that a square has a greater area than any other rectangle having the same perimeter. When we speak of "any other rectangle" we are referring to rectangles with smaller widths and greater lengths and rectangles of greater widths and smaller lengths. Should there be a finding that the subtraction of the area of any other rectangle from that of a square leaves a result which is greater than 0, then the point we have made about the unparalleled virility of the square in terms of its ability to cover area shall be proven. But we must first complete the solution of the inequality which gives us $10,000 - 200x + x^2 > 0$. If we factor this solution, we get the equation $(100 - x)^2 > 0$, which tells us that if x is not equal to 100, then $100 - x$ is not zero. As a result, we find that the square covers a greater area than any other rectangle having a similar perimeter.

But what if we are not very interested in constructing a building to grow mushrooms? What if we do not care about enclosing the lands upon which we allow our pet cows and horses to graze and thus have no need to calculate the lengths of fencing needed to prevent our beasts from wandering away and falling off a cliff? Is there any need for calculating the shapes that maximize a given area? One could vigorously shake one's head and declare that algebra is the handiwork of an evil force or one could be rational and consider the possibility that this type of calculation could have important applications for industry. Both the food and beverage industries use these types of calculations to determine how best to minimize the amounts of materials needed to wrap or cover a given item. This is not a very difficult point to understand because a bottling company, for example, does not want to use a 5-gallon jug to hold the amount of drink that would ordinarily fit inside a 16-oz. bottle. Even though some might cheer any bottling company with the audacity to try to stock such oversized containers on the shelves of the nation's super-markets, it would be extremely expensive and wasteful. Similarly, a cookie company that sells fig bars in steamer trunk sized boxes would find itself spending most of its profits on the packaging of its fig bars and comparatively little money on the ingredients. This is where the information we have about minimizing packaging materials is extremely useful. But we must also confess that our laborious discussion of area coverage by squares and other rectangles is not complete without pointing out that those who really wish to be efficient in their use of materials should consider the simple yet refined elegance of the circle.

Believe it or not, the circle is even more efficient than the square in covering areas. Despite popular demand, we will not spend the remainder of the book examining the algebraic proof which underlies this conclusion. But it should suffice to point out that there are many companies which have already caught on to this little secret about circles—namely those companies which use cans to hold products ranging from tuna fish to soft drinks. Companies that use cans want to minimize the amount of packaging materials needed to hold a given amount of product. Australian brewers, for example, are notorious for selling their beers in cans that are big enough to house entire families. This is not necessarily a reflection on the quality of Australian beers (which some snippy critics who do not appreciate such packaging have likened to "jug beers") but instead may be the practice of those who drink such beers to consume them in multiple-gallon quantities. Mathematicians are quite capable of proving that the cylinder is the most efficient container in terms of its capacity for enclosing the maximum amount of space with the minimum amount of material. What is surprising is that the utility of the cylinder has not been recognized by more persons and inspired the formation of cult-like groups which erect cylinder-shaped temples and collect money for cylinder research at airports and bus stations. That the beverage industry has settled on the cylinder as its shape of choice in the use of tin and aluminum cans is not a complete surprise because the more efficient design ensures the lowest possible packaging costs and the greatest profit. The cylinder can has many uses and may be used by persons with much time on their hands: You can cut the ends off a cylinder and make a telescope or you can use the can as a handy device to shatter your windows when you lock yourself out of your house. The basic principles of algebra and geometry are not easily contemplated when one is using one's telescope to spy on the neighbors cavorting in their hot tub but one can certainly appreciate their power to help us make decisions that maximize our resources and minimize our costs.

Algebra in particular is often singled out by disgruntled students as being boring or irrelevant because it strikes most of us as a foreign language. Well, it *is* a foreign language of sorts but it is fairly simple to learn if one has a little patience. Actually, one does not need to be supremely intelligent to master algebra because its logical structure is not that different from any spoken language. If you can master (or at least be familiar with) one language—whether it be English, French, Spanish, Italian, Russian, Chinese, or Japanese—then you can learn algebra. It is a type of language in which meaning and understanding are gained from the interpretation of its symbols. Of course algebra will not replace French as the language of choice for winning the affections of a loved one because very few ladies will swoon over the sight of their would-be suitor writing a quadratic equation in the sand below their window.

But algebra lacks the beauty and profound mystery of other disciplines such as physics or astronomy. It cannot claim to offer models that will enable us to understand better the lumpy distribution of stars in the universe nor can it give us insights into what sort of chemical processes on earth first gave rise to the most primitive forms of life on this planet. However, it can offer proofs for why cylinders are the preferred container for soft drinks or why the square should be used as the model for constructing windowless concrete greenhouses. Much of the lack of romance associated with algebra is that algebra students become so engrossed (or bogged down) in their efforts to "make sense" of it or to understand its relevance that they fail to see its primary function is calculations. Indeed, algebra may be thought of as something akin to a magnificent calculating machine although there is no obvious way to plug it in.

An understanding of algebra is important for students who want to move further along the superhighway of mathematics because algebra is usually the first branch of mathematics in which students begin to use unknown variables represented by such letters as x and y. Algebra thus entails an additional level of abstraction beyond that which is required of those who consider numbers in their basic form as with grade-school arithmetic. Only after students become comfortable with the idea of representing numbers with unknown variables can they advance into more complex areas of mathematics such as the calculus.

CHAPTER 6

To Infinity and Beyond

How quaint the ways of paradox—
At common sense she gaily mocks.
—W. S. GILBERT

The concept of infinity is one which has caused more than one mathematician and philosopher to go mad. It seems that one should not have to contort oneself trying to understand the difference between finite and infinite numbers. But it is a distinction which becomes more subtle and profound the more intently we consider it. If you were to go up to a person on the street who did not happen to have a doctorate in group theory or existential philosophy and ask her to discuss the differences between the finite and the infinite, she might fumble for words or try to sock you with her handbag. After being hit over the head a few times, you might decide it would be more prudent to read a book which explains the distinction between the two concepts if only to avoid a skull fracture.

Finite numbers include counting numbers such as 1, 2, and 3, which are used to describe groups of objects. Certainly we can see that the number 3 could be used to represent three fingers or three automobiles or three dogs. We could count the number of objects in each of these groups and verify that there are indeed three such items there. Of course if we are products of shaky reform school educational programs, we might come up with the wrong number. But these abstract numbers have a one-to-one correspondence with the objects being counted. The number 1 is assigned to the first object, the number 2 to the second object, and so on, until all of the objects have been so described. Because we can easily count the total number of objects in each of these groups, we can appreciate the fact that they are finite in number; we do not need to continue counting on and on and on and on—unless we are one of those feeble-minded individuals who loses his or her way when engaged in number crunching.

The concept of finite quantities appears in other ways. The earth itself is finite even though I could begin walking in one direction and continue my journey for all time without coming to an edge. Of course I would find myself crossing the surface of the planet over and over again but I would not have to worry about falling off the edge of the earth because it is a sphere. The warm feeling that comes with knowing such data would probably dampen as I continued trudging across arid deserts, steaming jungles, impassable mountains, spewing volcanos, and stinky swamps. But I would certainly have a wonderful opportunity to see much of the world without enlisting in the armed forces. The interesting thing about a sphere is that although it does have a limited circumference and diameter, I can walk across its surface forever without coming to an edge. Spheres are thus said to be finite but unbounded. In other words, a sphere such as our planet is finite even though I could theoretically walk an unlimited number of miles across its face. But the novelty of this activity will gradually disappear as I continue my travels— particularly when I start to see the same sights, such as the same volcanos erupting and burying hapless villages, the same deserts claiming the lives of travelers who happened to run out of gas in the wrong place at the wrong time, and the same impassable mountains gradually being turned into subdivisions.

Indeed, one might begin to wish that the earth was flat so that there was a logical stopping point to such wanderings. But perhaps it would be more helpful to focus on the idea of counting itself and the utilization of one-to-one correspondences between groups. This one-to-one correspondence is extremely useful when dealing with large numbers of objects or individuals and makes it possible to avoid counting every object that has a matched counterpart. This might be a little easier to understand if we think about the one-to-one correspondence that exists between the bottles and the bottlecaps for a given carbonated beverage, Asparagas, a lovely concoction of asparagus stalks and motor oil, which enjoys great popularity among vegetarian auto mechanics. For every bottle of Asparagas, there is a matching bottlecap. To determine the total number of bottles and bottlecaps, we do not need to count all of the members of each of the two groups but can instead count either all of the bottles or all of the bottlecaps. Because we know there is a one-to-one correspondence between the members of each of the two groups, we can simply multiply the sum of either group by two to get the total number of members in both groups.

Another example of this can be found if we visit the ballroom of the extremely snobby Von Ribb family where ladies and gentlemen are dancing the waltz to celebrate the anniversary of the Prussian invasion of France. If we ignore the few same-sex couples who have decided to use this occasion to "come out," we can assume that every pair of dancers consists of one female and one male. Therefore, there is a one-to-one correspondence between the number of ladies and the number of gentlemen on the dance floor. Once we have determined that this one-to-one correspondence holds true for the persons attending this ball, then it is a simple matter to count either the number of persons wearing dresses or the number of persons wearing tuxedos (assuming that all the cross-dressers have already been tossed off the premises by the security guards) to determine how many people are actually attending the ball.

We might also view the one-to-one correspondence in a way which everyone who has thrown a party can appreciate. Suppose that you are the parent of the birthday child and you are hosting a birthday party celebration for your precious dear at a park. Once the pizza delivery arrives and a horde of famished, screaming youngsters descends upon the serving table, you must quickly pass out one slice of pizza to each child or risk having a limb bitten off. If you are fortunate enough to have a few minutes' advance warning before the juvenile blitzkrieg occurs, you might fetch your paper plates and lay one slice on each plate, stopping only after you have enough plates out so that each child would have a single serving. The number of plates has a one-to-one correspondence with the number of children.

Because there is only one slice of pizza per plate, the number of children also has a similar one-to-one correspondence with the number of slices (or the number of plates). This is important because a misunderstanding of this idea of matching food with children could cause someone to pass out only a few plates, leaving the little people to engage in a bitter battle over who gets to eat pizza. Alternatively, one might order too much pizza to begin with, ignoring the obvious point that ten young children cannot eat 50 large pizzas in a single sitting, even if they are allowed to spit out the crusts and peel off the anchovies and vegetables. Here again, one can quickly determine how many children or how many slices of pizza have been passed out if we count the number of slices handed out and make sure that each child receives only one slice. Once we make sure that every child has a single slice (including those who have dropped their slices on the floor or thrown them at the chaperone), then we will know how many children we have in attendance.

This correspondence is necessary to understand more fully what we mean by the idea of counting. When we count the number of elements in a set such as the number of chocolate chips in a cookie baked by the Happy Baker Cookie Company, we use the class of integers (1, 2, 3, 4, 5, 6, and so on) and assign the 1 to the first chocolate chip, the 2 to the second, and so on, until we have assigned an integer to every chocolate chip. The last integer, which should be the largest integer if we have mastered basic counting, will tell us how many chocolate chips can be found in this cookie. If we engage in this operation from 1 to 16 before running out of chocolate chips, then we will have determined that the cookie itself contains 16 chocolate chips (unless, of course, we have already given in to our pangs of hunger and eaten the subject matter). The integer 16 denotes the total number of chocolate chips, thus giving us a snapshot view of the concentration of chocolate chips in the cookie. We do not need to keep recounting these chips to know whether it contains 13 or 18 or 15 chips. We already know that it is a 16-chip cookie, which meets the rigid standards imposed by the Happy Baker Cookie Company because it has the requisite number of chips needed for inclusion in a package of Choco-Chippies.

This idea of counting can be applied to all types of baked goods including cakes, cookies, and loaves of bread. Indeed, its applicability extends far beyond baked goods to items and objects of all kinds. But mathematicians would find our explanation of counting to be somewhat lacking in rigor and precision so we might want to express this concept in an abbreviated way that would have at least some of the semantic baggage of a textbook. Suppose that we have a class A which contains a finite number of elements. It is possible to find any number of other classes having the same finite number of elements so that each element in class A may be matched in a one-to-one correspondence with each of the elements in any of these other classes. Because each of these classes has the same number of elements (regardless of whether one class contains vermin and another class contains fluffy puppies), each is considered to be equivalent to the other (even though most people would prefer to wake up to the smiling chubby face of a fluffy puppy than the beady eyes and pointed noses of a bedful of rats). The idea of counting thus does not allow for subjective judgments as to the preferability or, indeed, quality of the elements of a class; it merely compares the quantity of elements within each class. Despite the differing nature of the elements in each class, however, they all share the same cardinal number, which is the integer we use to denote the total number of elements in each class. According to the mathematicians Kasner and Newman, the cardinal number of the class A is "thus seen to be the symbol representing the set of all classes that can be put into one-to-one correspondence" with A. If we have a number of classes having ten elements, then the number 10 is the cardinal number of each of those classes. Each of these classes thus has a cardinality of 10, regardless of whether the class consists of ten toes or ten ants or even ten honest politicians (although we might have to double count in the last class). So we can take comfort in the fact that the number 10, which appears at first to be a rather timid number, can represent the cardinality of any class having 10 elements with gusto and verve.

Although the definition offered in the preceding paragraph seems overly formal, it is nothing more than a recognition of the method which was originally devised by people to count groups of objects. It was not invented by mathematicians who suddenly decided that humanity needed to have some way to represent quantities of classes but instead preceded the first mathematicians by many centuries. Indeed, it is probable that the first counter was that person who saw that there was a common link between the five fingers on his hand and the five goats in his living room. Neither group had anything in common in an intrinsic sense but their quantities were in fact equivalent to each other. In the terminology of modern mathematics, one could say that the cardinality of the class of the fingers on one hand (assuming one had not severed any digits during a career as a near-sighted vegetable cutter in a fancy restaurant) was equal to the cardinality of the class of five goats bleating in the bedroom. In determining the number of objects in any class, one merely places the objects in a one-to-one correspondence with the class of integers as we mentioned above. This is an intuitive process which is used in everyday life by nearly every individual, nearly all of whom do not give this process a second thought. But it is universal in its application to any class consisting of any number of elements.

When we count the number of objects in a class, we seldom venture beyond a few dozen or even a hundred. It is simply too time-consuming and impractical for us to count all the elements in larger classes one by one. If I were to purchase an exotic automobile for $100,000, I would probably not bring 100,000 one-dollar bills to the dealer and painstakingly count them out onto his desk, one by one, because it would take hours, if not days, to finish this task. Even though automobile dealers are known around the world for their kindness and patience, it would be difficult to imagine one being so saintly that he or she would sit there with an unwavering smile as I counted the bills out ("Ten thousand forty-one, ten thousand forty-two, ten thousand and ... uh, oh, I lost my place.... One, two, three"). This practical difficulty of counting up to larger numbers underscores the importance of the concept of cardinality. The number 100,000 can be used to represent the cost of an exotic automobile or the face amount of a cashier's check. I do not have to bring 100,000 one-dollar bills to the dealer so long as I can bring a check that will cover the purchase price of the automobile. We know that the check for $100,000 is enough to pay for the automobile without having to worry about a one-to-one correspondence between the price of the automobile and the face value of the check. This is a very important consideration because one can only imagine the chaos that would occur if we lived in a world in which everything was paid for in one-dollar bills. Everyone would have to bring huge wheelbarrows of cash to the department store to purchase the newest designer clothes. They would probably have to bring entire carloads of cash to purchase a house or a business. One would become very used to the sight of people carrying huge sacks of money and the entire population would gradually develop bad posture from being weighed down with bags of cash. Fortunately, this sad state of affairs never came to pass because of the development of this idea of cardinality.

Anyone can count to a hundred or a thousand or even a hundred thousand if he or she has enough time on hand and the necessary patience. But as we deal with millions, billions, trillions, quadrillions, quintillions, sextillions, septillions, and even greater numbers where the zeros begin to flow like bad wine at a low-budget dinner theater performance, the willingness to count can flag. After all, what is the point of counting up to such big numbers, particularly when they seem to have very little relevance to our daily lives? The concept of cardinality makes such laborious exercises seem superfluous because we know we can use the symbol 1,000,000,000,000,000 to represent 1 quintillion without bothering to count every one of the 1 quintillion elements in the class. This saves a great deal of time and frees up one's evenings for the next 50 or so years so that one can enjoy life's pleasures instead of being cooped up in a dimly lit room hunched over a stack of papers, counting class objects.

A generation ago, we thought of a billion of anything as being virtually unlimited. But national budgets and astronomical distances have made the use of the word "billion" quite commonplace so that we no longer appear to be very intimidated by it. Nowadays, the word "trillion" as used in global commerce figures and currency flows is becoming a tired cliché. In another generation, we may become more acquainted with the next highest step in the numeration system—quadrillions. But this acquaintance will occur in the scientific world of subatomic particles and the celestial world of stars because it is unrealistic to expect that we will have a quadrillion people living on the planet or that automakers will produce a trillion vehicles. Astronomers and physicists, however, have long been accustomed to grappling with large numbers. Sir Arthur Eddington, the British astrophysicist, began one of his lectures with the declaration that there are 136×2^{256} protons and an equal number of electrons in the universe. Eddington's audience was so impressed by the preciseness of his number and the sheer audacity of his statement that they forgot to ask him how he had arrived at this number— which, in this case, was nothing more than an educated guess. Even though Eddington made his prediction at a time when nothing was known of the neutron or any of the other myriad elementary particles which have since infested high-energy physics, his statement underscored the power of numbers to describe in very little space very large quantities.

One of the more famous of the large but finite numbers is the googol, which owes its playful name to the nephew of the mathematician Edward Kasner and is represented by a 1 followed by 100 zeros. Dr. Kasner was not to be outdone by his nephew and he went on to coin an even larger number called the googolplex, which is 1 followed by a googol of zeros. One could go even further and propose the "magnagoogolplex" which is 1 followed by a googolplex of zeros or even, if we have a bout of mathematical megalomania, a "summagoogolplex" which is 1 followed by a magnagoogolplex of zeros. These numbers are unimaginably vast and would seem to have little practical application to our physical universe. Indeed, it is probably this lack of practical application that stopped mathematicians from bothering to continue naming ever higher denominations of numbers such as sextillions, septillions, octillions, and so on, because one would have very little reason to use such numbers in the real world. No doubt there is an untenured mathematician working in a dimly lit room on an obscure college campus gleefully creating new numbers followed by ever greater numbers of zeros. But even if we were to create numbers that would take a billion or a trillion or even a quadrillion years to write, these numbers would still be finite, albeit bigger than the proverbial breadbox.

The point of this longwinded discussion of big numbers is that anyone can create a huge number and name it after himself or herself if he or she is willing to write down enough zeros after a 1. But there is a certain unreality to such an exercise because one soon begins to wonder whether there is any point to such an exercise. After all, it is very exciting to create a new number like the "spudillion" which is 1 followed by 933 zeros. But what good is the "spudillion" (which has been named to honor the potato) if one never has any occasion to use it? You cannot go into the snobbiest French restaurant in town and ask for a spudillion glasses of wine because such a demand would clearly outstrip the global wine supply for at least the next 5 billion years. Of course you would have ample time to drink yourself into oblivion and would probably not care about the inability of the restaurant to fulfill your request. But it underscores the point that numbers that have no relation to the physical universe or our world of everyday affairs become almost ghostlike.

At the same time, we have had to grapple with numbers which would have been unthinkably vast to earlier generations. Fortunately, we have an elegant and powerful system of mathematical notation which makes it possible to express extremely large numbers in very compact spaces. This is a very remarkable quality that permits us to conceptualize and simplify very large quantities that would otherwise be very cumbersome and clumsy to write. Our use of and comfort with large numbers is comparatively new because the thinkers of antiquity, subject to a few exceptions, were not very interested in developing, let alone using, large numbers. One reason for this reluctance was the absence of a convenient numbering system. Although the Romans had made some advances over earlier numeration systems—such as those used by the Greeks—by making limited use of positional notation, the Roman system of numeration still had its drawbacks. In essence, it incorporated a mix of letters such as X, V, C, and M which were used to describe the numbers 10, 5, 100, and 1000, respectively. One would then arrange the letters to create the desired number. MM would be equal to 2000 whereas CXV would be equal to 115. The problems with this system became more acute as one attempted to create steadily greater numbers—due in part to the limited number of alphabetic characters and the fact that multiple characters (such as VIII for the number 8) would have to be used to represent a single number, thus compounding the unwieldiness of the numeration system. The Romans, who were not known for their sense of humor or brevity, were not the best persons to develop an improved numeration system. They were, after all, busy conquering and killing people in foreign lands because one cannot tolerate people who wear unfamiliar garments and speak strange languages and eat exotic foods in one's vassal states. As a result, the Romans were very serious about whatever they set out to do; they did not have the oomph to develop a zippy numeration system which could incorporate such things as exponents to represent extremely large numbers in very short spaces. If one wished to write out the Roman version of a number such as 856,885,114,325,967,878,321,986,933,567,002, one could expect to develop a severe case of writer's cramp by the time enough C's, M's, X's, V's, and I's were strung together.

This is not to say that the ancients were completely oblivious to large numbers. Many of the early astronomers believed the number of stars in the sky to be infinite in number even though they could only see a few thousand on the clearest, darkest nights. But the Greek mathematician Archimedes realized that no matter how big a finite number, one could always create an even bigger number by adding 1 to it. He was also keenly aware of the tendency of most persons to confuse big finite numbers with the infinite and in his work *The Sand Reckoner* he pointed out to King Gelon that the number of grains of sand in the world, no matter how great, is still finite because the world itself is finite. Archimedes was one of the first to recognize that finite and infinite classes are fundamentally distinct from each other in that the former does not at some point become the latter. In other words, if I begin counting as fast as I can, day in and day out, week in and week out, there will not occur some magical moment in the 936th week whereby I reach infinity in my counting. Indeed, I could count for ten thousand years and be no closer to infinity. Now this might strike you as very strange because surely ten thousand years is enough time to count to infinity. But the fact is that there is a yawning chasm that separates the biggest finite number we can think of from the infinite—one that is qualitative as much as it is quantitative. Finite numbers are elements in the class of numbers which we call integers and this class is in turn infinite. No matter how long we count (whether we hold our breath or not) we can never reach the end of the class of integers because we can always create a new integer by adding 1 to our highest number.

So why is there this concern or even, some would say, this unhealthy fascination with the infinite? We do not really deal with infinite quantities in day-to-day life. But this concept perplexes us because the fact that we must deal with finite quantities necessarily requires us to consider the infinite. We have all had experiences where we begin to debate with the persons sitting around us deep philosophical points such as whether there are actual boundaries to the universe or whether there is such a thing as a single biggest number or even whether Goldilocks was prejudiced against bears. In these situations, one can usually expect to engage in stimulating conversations with sharp-witted souls who believe, for example, that houseplants are the descendants of a long-vanished alien master race which visited the earth many thousands of years ago or that you can hear the launch codes for the nation's intercontinental ballistic missiles in a bowl of crackling rice cereal if you listen closely enough. But these experiences do occasionally foster a sense of introspection, due in part to the difficulty of focusing one's eyes on any external objects. And when this newfound ability causes one to think about the dichotomy between the finite and the infinite, the reasoning process starts to break down. Many people think of the infinite as the last number in a sequence of numbers, thus implying that one can reach the infinite merely by counting long enough. Others seem to understand that the infinite is different from the finite but they grasp for words to express that difference.

The infinite should be easy to understand because anything infinite is simply that which is unlimited or unending. A conversation with a spouse in which your own shortcomings are explained in great detail can seem infinite in duration but it will eventually end if only to allow for the taking of meals or sleep. But we all have problems trying to imagine something that never ends because we are so accustomed to thinking in terms of the points of reference which abound in our own world. To consider an infinite physical universe is difficult because our own experiences with spatial distances such as lengths or widths or depths involve compact areas which are typically limited or bounded in some way. When we are using a map to help us determine when to turn off or turn on to a given road, we are necessarily dealing with a finite length of highway which we know will eventually end, despite the seemingly interminable battles being waged by the children in the back seat. But how would we feel if we knew that this road would never end—that we would continue following it for hundreds, thousands, millions, and even billions of years into the future with no end in sight? Well, we would want to make sure that we had a full tank of gas before beginning our journey.

But our previous venture in the creation of ever greater numbers illustrates a process of recurrent reasoning whereby we are able to construct numbers as large as we wish without actually having to count one by one to that very same number. In this way the process by which we create ever-greater numbers becomes apparent but we do not have to engage in the laborious process of trudging from one number to another. We thus know that we can approach the infinite and actually comprehend what it means to be infinite even though we cannot directly experience it. Of course this is not a perfect guide but has an air of unreality about it—much like trying to understand the nature and character of a woman merely by looking at the pictures of many women.

The Paradoxes of Zeno

Because many clever people have debated the intrinsic characteristics of the infinite for thousands of years, it should not surprise us that the same points and arguments have resurfaced over and over again. One of the earliest persons to immerse himself in the intricacies of the infinite was the Greek philosopher Zeno who offered a number of paradoxes to aggravate his contemporaries and torture future generations of philosophers. Because Zeno did not have a regular job which demanded all his attention, he spent much of his time devising elaborate proofs which demonstrated, for example, that motion is impossible. Such a proposition might surprise persons who experience motion on a day-to-day basis when walking from their house to their garage or down the aisles of the grocery store. But Zeno was a master of clever reasoning and he demonstrated this in his paradoxes even though the correctness of these paradoxes has been the subject of vigorous, indeed, fanatical disputes. As far as proving the impossibility of motion was concerned, Zeno offered the following explanation: If one was to traverse a given distance (perhaps from the bathroom to the guest room with the mirrored ceiling and the revolving coin-operated bed), one would first cover one-half of the distance. One would then traverse one-half of the remaining distance and so on and so forth without limit. Each such interval would be half the size of the interval preceding it but there would be an infinite number of such intervals in the distance separating the two rooms. As such, it would seem that one would never make it to the mirrored guest room in one's own lifetime, let alone before the time expired on the spinning bed. But there is a flaw in Zeno's paradox which is the finite nature of this series. This infinite series of terms does not in turn constitute an infinite space as assumed by Zeno—it instead constitutes a finite distance. In this case, the distance is that separating our bathroom from our guest room. It does not take us an infinite period of time to make it from one room to the other, even when we have become very disoriented after a night sampling the fine wines of Upper Volta at a local wine tasting party. We may stumble hither and thither but we will eventually navigate that finite distance. Zeno assumed, however, that an infinite number of intervals necessarily entailed an infinite distance, which, of course, cannot be traversed in a finite period of time.

Always the life of the party, Zeno also offered a paradox in which he proposed that a tortoise, when given a head start in a race, could never be overtaken by the fleet-footed Achilles. The reasoning was similar to that described previously, in that the tortoise, plodding ahead at a breakneck speed of one-tenth of one mile per hour, would reach a certain point, say, 30 feet ahead of the starting line. At that point Achilles would lace up his running sandals and take off. But by the time Achilles reached the 30-foot marker, the tortoise would have lunged ahead to the 32-foot marker. No doubt Achilles would gleefully have the tortoise in his eyes as he raced ahead. But Zeno's paradox would seem to pose an insurmountable problem to Achilles' attempt to catch up to the tortoise because by the time

Achilles reached the 32-foot marker, the tortoise will have reached the 32⅛-foot marker. So no matter how fast Achilles ran, he would always be behind the tortoise. Or at least this is the conclusion that Zeno's paradox would appear to lead one to embrace. Of course Achilles might not be familiar with the irrefutable logic offered by Zeno and simply race past the tortoise. But if we accept Zeno's premises, then it appears that Achilles could never catch the tortoise.

Zeno also pointed to the example of an arrow in flight to prove that motion is impossible. At every instant in its flight the arrow is at a single point in space. But because it must always be at some point in space, it cannot be in an interval between any two points and thus cannot travel between points. If it were to travel between points, it would be in a non-point which means that the arrow would be nowhere. Now even though Zeno was not known to indulge in hallucinogenic substances, the arrow paradox does seem to suggest that Zeno had become more enamored with the impregnability of his terminology than with the reality of an arrow in flight. It is not known whether any of Zeno's students challenged him to stand in a courtyard with an apple on his head to test out the apparent impossibility of motion. But perhaps Zeno would have recognized that even the best of theories sometimes run aground on the cruel rocks of practical reality.

Even though these paradoxes do not appear to hold true, they caused many of history's greatest philosophers to wander into semantic swamps and logical morasses and debate their merit. Aristotle himself became hopelessly tangled up in their intricacies and was forced to move on to more fruitful areas of inquiry. Many of Aristotle's successors were forced to draw the same conclusion and leave it for their own unsuspecting successors to grapple with. Indeed, it was not until the 19th century that several European scholars revisited the concept of infinity and first considered it in a mathematical as opposed to a metaphysical sense. The person most responsible for this shift in emphasis was the mathematician Bolzano who authored a concise treatise on the subject which was notable both for its stark prose and its disregard of the theological baggage which had long dogged the concept of the infinite. Bolzano's work provided a much-needed kick in the pants for the mathematical community which had up to this time been content to ignore the infinite, having happily consigned such issues to the theologians who were still arguing over the number of angels that could dance on the head of a pin.

Cantor's Battle with the Infinite

It fell to Georg Cantor, who entered the world in 1845 in St. Petersburg, to grab the concept of the infinite by the horns and wrestle it to the ground. Georg was the son of a merchant by the same name and the former Maria Bohm. He was a young child when Bolzano's book first appeared and, after spending his early childhood in Russia, his family emigrated to Germany. Young Georg had a passion for learning, particularly mathematics, which manifested at an early age. In the modern language, Georg was something of a "geek" but he was confident in his scholastic abilities. Even as a young man Georg became fixated with the fine points of Christian theology and this interest may have first sparked a desire in him to better understand the infinite.

Despite Bolzano's work, the infinite was still more a prisoner of theology than a recognized mathematical concept. It owed its theological significance to the inability of scholars to comprehend a physical infinity except to imagine a cosmos governed by the handiwork of a divine being. Even if one concluded that the universe was finite, then the inevitable question would follow as to what could possibly lie beyond the edge of the universe. At that point the discussion would degenerate into murky, often bitter, discussions which could not be resolved, thus leaving everyone to withdraw to the safety of his own prejudices.

But it was by no means assured that Georg would be permitted to pursue a mathematical career. His father had other ideas about his son's future career and urged Georg to study engineering, going so far as to suggest that it was God's will that young Georg devote himself to a more practical profession. Although Georg initially tried to accommodate his father's wishes, his enthusiasm was feigned and his efforts half-hearted. But he had such an intuitive gift for mathematics that he could have been a first-rate engineer even though he would have preferred to grapple with the most painful mathematical proofs than to design a bridge or a road. Yet Georg's father eventually gave in to his son's ambitions to study mathematics, permitting him to enter Zurich in 1862 and then, a year later, the University of Berlin. He completed his doctoral dissertation in 1867, which dealt with the solution for an indeterminate equation first considered by Gauss. Cantor's work was outstanding but it was also conservative and gave no indication that his later work in the infinite would shake the very foundations of mathematics.

Cantor would never realize his dream of being a professor at the University of Berlin but was cursed by fate to spend the most productive years of his career at the University of Halle, which we might liken nowadays to a low-end community college. Unfortunately for Cantor, he began his teaching career as a tutor; he was paid not by the university but instead by the students whom he taught. Although such an idea might seem absurd to modern students who can barely rouse themselves to get up and go to their morning classes, it was a common practice in 19th-century Europe, although one must wonder how successful Cantor was in collecting fees from students whose spending priorities usually lay more with entertainment, sports, and various beverages. Cantor did eventually become a professor at Halle in 1879 but by this time he had completed his most important work in mathematics. This work was one of the most remarkable and original creations in the history of mathematics but Cantor had the misfortune of spending his career in the nation that possessed perhaps the single most reactionary mathematical establishment in Europe, one that was clearly not predisposed to consider the validity of his ideas.

In dealing with the infinite, one has to be willing to forget many of the rules one has learned in dealing with finite numbers. This is not a difficult task for many students who manage to forget most of the details of any course in mathematics soon after the final exam has been completed. But the mathematician who deals with infinite numbers must be willing to consider rules that would be absolutely nonsensical in dealing with finite numbers. First and foremost is the time-honored principle that the whole is greater than any of its parts. Even those persons who watch network television programs regularly can grasp this point very quickly. If we deal with the set of integers consisting of those numbers including 1, 2, 3, 4, 5, 6, 7, 8, 9, and 10, we can see that this class of integers constitutes a finite set. If we were to add up all ten of these numbers, we would obtain the quantity 55, which is clearly much greater than any single or multiple combination of these numbers. This is clearly an example of how the "new math" can bring untold riches to our humdrum, dreary lives. But it also shows that the arithmetic of finite quantities is intuitive because we would certainly expect to see such a result in our day-to-day experience with these numbers.

But what about those tricky infinite numbers that so bedeviled the early Greeks that many of them could not figure out their sexual orientation? The person considering infinite classes must accept the seemingly unbelievable statement that the whole of any infinite class is no greater than some of its parts. In other words, an infinite class may contain certain numbers which in turn possess an infinite number of elements. This statement sounds as though it belongs in an introductory philosophy course but it is clearly within the realm of mathematics as was shown by Cantor in a series of important papers completed during the fourth decade of his life.

Before we continue with the story of how Cantor's own revolutionary work caused him to become mired in a lifelong battle with several of his more ossified mentors at Berlin over his work on infinite sets, we need to explain what we are referring to by the statement that infinite quantities contained within an infinite set can also have infinite sets of elements. The concept of one-to-one correspondence can help us to understand this difference between finite and infinite sets. For example, we might imagine that we have a set consisting of all the ditch diggers in England who have completed doctoral dissertations on 18th-century French poetry. This would probably be a very small class of ditchdiggers although one can imagine their afternoon tea parties would consist of very provocative criticisms of the leading literary luminaries of that era. Certainly this class would be finite and would undoubtedly be smaller than the class of ditchdiggers as a whole or even those ditchdiggers who only completed their undergraduate education. Both of these classes are finite as we do not have an unlimited supply of ditchdiggers, regardless of their affinity for French poetry. If we have enough time, we can count the number of ditchdiggers in each class, which is, ultimately, the difference between finite and infinite classes. Here we are dealing with two finite classes, the larger one of which is the group of ditchdiggers that has not yet had the fortune to acquire a taste for cutting-edge French poetry such as *Frère Jacques*.

Both of these classes of ditchdiggers can be placed in a one-to-one correspondence with each other. If there were 38 ditchdiggers who had contributed to our knowledge of French literature in their doctoral dissertations, they would clearly fit within the larger class of 356 ditchdiggers as a whole. The 38 ditchdiggers would constitute a unique subclass which could be placed in a one-to-one correspondence with the larger group of 318 ditchdiggers who have specialized in other areas of study such as high-energy physics and metaphysical logic. We would then be able to determine that by placing the two classes in a one-to-one correspondence with each other there would be an extra 280 ditchdiggers left over. Because the larger set has a finite number of additional elements, it is necessarily finite.

Why would anyone want to travel to the far corners of the world when they could spend their days on the beach drawing diagrams of finite classes in the sand? As we have probably explored the one-to-one correspondence of elements in finite sets in enough detail to satisfy even the Marquis de Sade, we must now shift to something which is much less intuitive and, hence, much more tantalizing—the one-to-one correspondence of elements in infinite sets.

But as we want to plunge ahead and thoroughly immerse ourselves in the concept of the infinite, we need to examine the way in which we determine the relative numerosities of two infinite sets. When we deal with infinite sets, we may have elements of infinite sets which are themselves equivalent in cardinality (or numerosity) to the infinite set itself. In other words, some of the parts of the infinite set may be equal to the infinite set itself. But how can this be? you might ask, if you truly cared about the plight of the human condition. Let us take a very simple example. The class of integers, which consists of the numbers 1, 2, 3, 4,... and so on without end, is an infinite set. By this characterization, we are saying that we can begin counting at 1 and continue counting forever without reaching the end of this set. No doubt there is a government bureaucracy in which highly trained numerologists spend days and days counting infinite sets, not terribly concerned with the fact that they will never complete their task (a situation which guarantees a certain level of job security). But we do not have days and weeks to spend in mind-numbing and, ultimately, pointless counting exercises to educate ourselves about the intricacies and nuances of the infinite. We can simply make the statement that the set of integers which consists of the numbers 1, 2, 3,... and the seemingly smaller set of integers which are divisible by 100 (100, 200, 300,...) are equivalent. In other words, both sets have the same number of elements even though one set would, at first glance, appear to have 100 times as many elements as the other. Mathematicians are able to prove this equivalence in a very simple way by placing the two sets in a one-to-one correspondence with each other. Here, 1 corresponds with 100, 2 corresponds with 200, 3 corresponds with 300, and so on. Because both sets continue onward without limit, they are both infinite and, as a result, are both equivalent to each other. The second set (the numbers divisible by 100), which is certainly a subset of the first (the class of integers), is a part of the class of integers but still has the same numerosity as the set of integers. Both are infinite which means that both are equal to each other. This is not merely the babble of a holy man who has been sitting in the desert sun for too many months but a striking example of the allure of the mathematics of the infinite.

This is not an isolated incident in the shadowy world of infinite classes. Indeed, there are many other classes of integers which can be placed in a one-to-one correspondence with the class of integers itself and still have the same cardinality. We could, for example, write down the set of integers divisible by 300 (300, 600, 900,...) and we would similarly find that there are as many such integers as there are integers which are divisible by 1. This is really quite an extraordinary feat, far more impressive than pulling a rabid rabbit out of a magician's hat or making a government tax auditor disappear. The fact that we can place these sets of numbers in a one-to-one correspondence with each other is also a handy time-saving device because we simply could not live long enough to count the elements in these sets. So if we know that the set of integers (1, 2, 3,...) is infinite and that we can place each of the numbers of another set (such as the set of integers divisible by 300) into a one-to-one correspondence with it, then we know without even having to attempt a futile counting task that the two sets are equivalent and, hence, infinite. This underscores the fundamental truth of infinite sets which distinguishes them from finite sets of numbers—the fact that sets which would appear at first glance to have far fewer numbers actually have the same (unlimited) number of elements as the class of integers itself. Indeed, there are an infinite number of classes which can be extracted from the class of integers; it makes no difference whether we use every tenth, three hundredth, four thousandth, six millionth, two billionth, or twelve trillionth number. All are equal in numerosity to the class of integers.

It is the placing of these infinite classes into a one-to-one correspondence with the class of integers which makes it possible for them to be "counted." In this case, we are pairing up the members of the class of integers (1, 2, 3,...) with the members of a particular subclass of integers (300, 600, 900,...) with the same grace as the mashing together of couples at a high school dance. But the mathematical operation is much less aggravating as we do not have to worry about monitoring sweaty groping sessions in the outside parking lot. It also enables us to assume that both classes continue without limit which frees us from the responsibility of having to count the numbers in each class. We can thus count both classes in an instant because we know that both classes are infinite.

How can one express the cardinality or numerosity of infinite classes? One can say that the cardinality of the class of integers is infinite but that does not sit well with mathematicians who wish to be more precise and create a unique terminology to express the cardinality of infinite classes. Indeed, Cantor was concerned that the mushiness of the term "infinite" would frustrate efforts to understand better the concept and so he thought it preferable to create a new mathematical term to express the cardinality of the class of integers. As he did not feel it would make much sense to select an integer to express the cardinality of an infinite class, he decided to use the Hebrew letter "aleph," which is represented by the symbol $\aleph$. This choice was completely arbitrary as Cantor could have used any letter of our own alphabet or even a drawing of a radish. But Cantor thought that a distinctly nonmathematical symbol such as $\aleph$ was called for because he did not want to have his term confused with an algebraic symbol or a cookbook recipe or even a lingerie catalog. However, Cantor suspected that there might be more than one type of infinite class so he assigned the term "aleph null" ($\aleph_0$) to describe the cardinality of the class of integers. This is an example of what is called a "transfinite" number, a term which describes the numerosity or cardinality of an infinite set.

Even though Cantor was prone to bouts of madness in his later years, his reasoning for the creation of a complete hierarchy of transfinite numbers was sound. He proposed that there is an infinite number of transfinite numbers, beginning with aleph-null and continuing with aleph-one ($\aleph_1$), aleph-two ($\aleph_2$), aleph-three ($\aleph_3$), and so on. One could, therefore, construct transfinite numbers of any cardinality, which would each have varying degrees of "numerical density" but all of which would themselves be infinite. By "numerical density" we are referring to the difference that exists between the set of integers $(1, 2, 3, \ldots)$, for example, and the set of integers divisible by 100 $(100, 200, 300, \ldots)$. Both are infinite but the set of integers is a "denser" infinite set than the set having numbers which are multiples of 100 because a new element is created every time we add 1 to the previous number. In the second set of integers, by contrast, we can create a new element only by adding 100 to the previous number. The first set thus gives us 100 new integers for every 100 places whereas the second set gives us only one new integer for every 100 places. In this way, we are able to make the seemingly paradoxical statement that both sets have an infinite number of elements but that one set is 100 times more dense than the other.

Having become acquainted with the concept of transfinite numbers, you are now ready to enter the exciting world of jet engine repair and make big dollars in your spare time. However, you will have to learn a few more things about infinite sets before you can take a wrench in hand and begin banging on the engines of a commercial jetliner. To begin with, you need to splash some water on your face and remind yourself that the transfinite numbers are no different in certain respects from regular counting numbers. They are governed by many of the same rules of logic as countable numbers and can be said to represent the logical extension of basic mathematical processes. But the transfinites are a sneaky bunch because they are affected in different ways than finite numbers by basic mathematical operations.

This distinction occurs because the transfinite numbers are themselves a strange breed of numbers. That seemingly innocent looking aleph-null holds the entire set of integers, which is an unimaginably vast collection of numbers that we would have no hope of writing out unless we had infinite time and a piece of paper that stretched outward to infinity in all directions. Because putting numbers on paper for all eternity is not what most parents would consider to be a prudent career choice for their children, we are fortunate in that we can use the transfinite aleph-null to represent the set of integers in our mathematical operations. Although many readers would doubt the practicality of this knowledge, one can readily see how it might be very useful in a volatile situation in which another driver has nearly run you off the road and you and he are now trading heated insults, trying to top each other in terms of the vulgarity of the particular expression. Your nemesis might tell you to go do something to yourself which is biologically impossible 10 times or 15 times. But you, having schooled yourself in the intricacies of transfinite numbers, can go tell him to do something to himself which is biologically impossible aleph-null times. Unless your adversary knows anything about the hierarchy of transfinite numbers, you will get the last laugh because there is no integer which is not contained within the aleph-null set. Consequently, your insult will be greater and have the added advantage that your nemesis will not really know what you are talking about. He will then be less inclined to start emptying the rounds of an automatic weapon in your direction.

To appreciate the wacky world of transfinite numbers, we need to look at aleph-null more carefully and see how it fares when we use it in basic mathematical operations. Suppose we have just finished lunch and decide that we wish to add aleph-null to 54. What do we get? Aleph-null, of course. After all, we already know that 54 is an integer which is contained within the set of integers. What if we add aleph-null and a googolplex or even our summagoogolplex? In either case, we are still left with aleph-null. Indeed, we can add aleph-null to itself ($\aleph_0 + \aleph_0$) and still be left with aleph-null. You see there is a certain wondrous aspect to this operation in that we always end up with aleph-null, making it unnecessary to remember tedious drills and freeing up much time to watch violent television programs.

What happens when we multiply aleph-null by another number? Aleph-null multiplied by 1 is aleph-null. Aleph-null multiplied by 2 or 10 or 80 or 6000 or 12,000,000 is aleph-null. In fact, aleph-null multiplied by itself is aleph-null. This monotony is wonderful in that we can, with the snap of a finger, master the entire addition and multiplication tables for aleph-null. This is a skill that not everyone can claim and it can be used to take intellectual snobs down a peg or two because this is the type of knowledge that sounds impressive while being very obscure.

But the monotony of the transfinite number aleph-null does not hold true for all mathematical operations. If we raise aleph-null to the aleph-null power, we get a new transfinite number called aleph-one ($\aleph_1$). We need to point out that this was not some sort of mathematical sleight of hand or merely a mindless bit of symbolic manipulation. Instead it came about through Cantor's own laborious efforts to prove that there are infinite classes which have a cardinality greater than aleph-null, one of which is the set of real numbers consisting of all rational fractions (e.g., ½, ⅔, ¾,…) and irrational numbers (e.g., $\sqrt{2}, \sqrt{3}, \sqrt{4},…$). The cardinality of this transfinite number, which is known as the cardinality of the continuum and represented by "C," is believed to be equal to aleph-one but mathematicians have not yet been able to prove this hypothesis. As with aleph-null, the arithmetic operations associated with C are strangely familiar but here the beast of prey is not a finite number but instead aleph-null itself. Whereas before we saw that the ravenous aleph-null could gobble up any finite number and still remain aleph-null, we now see that C (or, as we suspect, aleph-one) has a comparatively insatiable appetite and can swallow aleph-null without incurring a bout of indigestion. We can add 50 aleph-nulls or 500 aleph-nulls to C and we still get C. Thus one can go up to the side of a building armed with a can of spray paint and add any number of aleph-nulls to C and still get C.

The same rules hold true for multiplication, which means that we can get a bucket of whitewash and a roller and paint all sorts of equations which show that C multiplied by aleph-null or 50 aleph-nulls or even 500 aleph-nulls is still equal to C. Of course you may have to stand on precariously stacked arrangements of boxes to reach high enough to truly drive this point home with enough equations but your heart will be filled with joy that you will be educating the wider ignorant lay audience in the mysteries of transfinite numbers. This is a very important point to make when you are arrested by the police for defacing private property and must spend an evening in the local jail because you spent too much money on rollers and paint to post bond. The bright side of this situation is that jails are often filled with persons who know very little about Georg Cantor's work or even the details of his turgid mathematical proofs and represent a "captive" audience to which one may give extensive lectures, at least until being felled by an errant knife. But most jails are located near top-notch hospitals and knife wounds are usually not fatal. So you have a better than even chance of survival unless, of course, you persist in scrawling convoluted mathematical equations on the jail floor with your fingernails and cause the less stable inmates to finally lose control and descend upon you like a pack of rabid dogs.

Our experiences with C mirror those with aleph-null in that C appears to swallow up the lower infinities without so much as a belch. But Cantor demonstrated that we can create a new transfinite number by raising C to its own power, thus giving birth to another transfinite number, which is represented by the letter F. As with C, mathematicians suspect that F is the equivalent of aleph-two but have not been able to demonstrate the truth of this assumption.

When considering classes with the cardinality of C (or aleph-one), we use the class of real numbers—instead of the class of integers—as our basis for analysis. This means that the class of real numbers will serve as a sort of measuring stick for any classes having a cardinality of C. You will be relieved to know that one such class having a cardinality of C is the points on a line segment. We know, for example, that there are an infinite number of points on a given line segment, regardless of its length. Philosophers and other people who have time to attempt to count such things could draw a line in the sand and start counting all of the points on the line but they would soon discover that they could always place a new point in between any existing two points on the line. After a while, the futility of trying to count these points would become apparent to all but the most thickheaded of individuals and they would abandon the actual task of counting. It would also become increasingly difficult to place newer points between the existing points without marring the line itself. At this stage we would have to take it as an article of faith that one could, in theory, continue to place points between existing points without limit. This conclusion would be based upon the premise that there are an infinity of points between any two points on a line segment. Mathematicians refer to lines as being "everywhere dense" to account for the fact that no matter how closely together we locate two points, we can always dredge up an infinite number of other points to place between these two points. Cantor showed that the points on the line can be paired with the class of real numbers so that one could then assign the transfinite C to represent the cardinality of the points on any given line. What many people find so perplexing about the cardinality C is that there is no difference in the number of points on a line that is 1 inch in length or a line that is one light-year in length. Of course we cannot prove this statement directly by laying a 1-inch long piece of string next to a string that is one light-year in length because it is very time-consuming to unroll a string that is one light-year in length and then mark it up with an infinite number of points between every two points. But both strings are "everywhere dense" and have the same cardinality C. We could entertain ourselves by constructing an elaborate geometric proof to show that both line segments have the same cardinality due to the one-to-one correspondence of each to the class of real numbers but we will restrain ourselves as it is unhealthy to have so much fun.

Before leaving the idea of the cardinality of infinite classes altogether, we should point out that the cardinality of the entire three-dimensional space which contains our universe is equivalent to the cardinality of the 1-inch piece of string. This seemingly astonishing statement should not surprise us in view of the rules which govern transfinite numbers. But this three-dimensional space is really no different than a two-dimensional line because both are "everywhere dense." This statement may appear paradoxical to those untutored in the ways of Cantor's elaborate mathematics but it is of no consequence to those of us who flit about the terrain of mathematics like a moth circling a light. We merely need to recall that many of the parts are equal to the whole and we shall avoid those concerns grounded in the logic of finite mathematics, the very same logic which seemed to disappear once we first began wading into the swamp of transfinite numbers.

Given the queerness of the transfinite numbers, it should not surprise us that Cantor himself had a very rocky career as a mathematician. His ambitions to teach at Berlin were forever stymied by the hostility of his mentor Kronecker, who disagreed with the fundamental principles of Cantor's work. Cantor himself was more successful in self-propagation, marrying the former Vally Guttmann and fathering six children, none of whom apparently shared their father's enthusiasm for transfinite mathematics. He also struck up a friendship with Dedekind, who was a profoundly original thinker destined, like Cantor, to be mired in positions at mediocre institutions throughout his career.

Cantor's important contributions to the study of infinite numbers essentially ended in 1884 when he suffered a nervous breakdown. He would be tormented by recurrent bouts of depression for the remainder of his life and became well acquainted with the inside walls of various mental institutions. Although his supporters grew more numerous with time, it was not until the end of his own life in 1918 that Cantor achieved the recognition and honor which had eluded him for so much of his career. By this time, his work on the infinite had become an accepted part of the body of mathematics even though there still remained a number of inconsistencies and errors in his work.

The story of Georg Cantor and his work illustrates a sad but true fact of life in the sciences. Those persons who dare bring forth radically different ideas and theories are not always welcomed as conquering heroes. Indeed, the reaction of the establishment may be one of cold indifference or even outright hostility as was the case with Cantor. The sciences are rooted in a tradition of evolutionary progress, boosted by occasional revolutions and burdened by the prejudices of the past. Knowledge thus develops in fits and starts, careening forward at times and sinking backward into the morass of conservativism and orthodoxy.

By "conservativism" we are referring to the eminently sensible tendency of scientists to favor that which is established and proven as opposed to that which is speculative and without evidence. This conservatism has proved to be an invaluable aid to those creative persons with new ideas who have forced themselves to fit their brilliant insights into the boundaries of existing theories. If the idea, no matter how wonderful or original, could not be reconciled with the existing orthodoxy, then the idea was usually put aside by mainstream science. By the same token, however, once the orthodoxy had evolved to a point where the once discarded idea could be accommodated, then the idea itself might be dusted off, perhaps recast in new clothing, and incorporated into the orthodoxy. But such a practice does have a price in that it can cripple the acceptance of a new mathematical or scientific breakthrough such as Cantor's work in mathematics. Fortunately, science as a whole has a fairly good record of retrieving or resurrecting that which has been unjustifiably cast on the dust heap of history's ideas. The problem with relying on this mechanism is that it can take decades or even centuries before a worthy idea or theory may be retrieved and re-examined with the critical, objective eye that it merits. These time lags are of scant comfort to the creators of these ideas because innovation often exacts a toll which may leave these free thinkers toiling in obscure institutions throughout their careers, seeing unworthy rivals rewarded for comparatively worthless achievements, and encountering ridicule and even outright hostility from the scholarly community. Although Cantor was forced to suffer for his work, he did at least live to see most of his contemporaries accept his work as a fundamental contribution to mathematics. In a sense, he was able to have the last laugh, albeit a hollow one.

CHAPTER 7

Measuring and Designing the Heavens

All the inventions that the world contains,
Were not by reason first found out, nor brains;
But pass for theirs who had the luck to light
Upon them by mistake or oversight.
—SAMUEL BUTLER

Much of the development in mathematics has been prompted by practical considerations such as the measurement of land and the counting of livestock. But the early mathematicians also had something more in mind than simply tallying the number of sheep running around their backyards. Indeed, they were curious about the world around them and sought to use their basic knowledge of mathematics and astronomy to impose some type of order upon the seemingly capricious actions of the observable universe.

There is no single point in human history such as October 7, 2002 B.C. (early afternoon) when humanity first looked up into the evening sky and wondered about the swirl of stars circling overhead or the moon moving across the sky. Because of the primitive state of human knowledge at the time and the absence of any precise scientific instruments such as telescopes, these early astronomers had no way of studying the endless canopy above the earth with any precision. As they were also lacking for entertainment, many of these celestial voyeurs invented stories and even imagined certain patterns of stars to be shaped as objects which make up what we now know as the zodiac. At the time they did not know that they were laying the groundwork for opening lines which could be used in taverns to entice the fairer sex such as "What's your sign?" or "My biorhythms are throbbing for you." But the important point is that there was little concern with actually verifying the scientific basis for such ideas at this time because science and the scientific method, as we know it, did not exist—at least not in its formal state. No one was very concerned about formulating rigorous theories about the external world and then testing them against the observed evidence because that was not as much fun as superimposing imaginary pictures on the stars of the constellations.

Certainly these ancient astronomers had no inkling of the role mathematics would play in the development of the embryonic field of astronomy, which was destined to be inextricably intertwined with astrology until the time of Galileo. In particular, these early celestial observers could not have in their wildest dreams anticipated the role that the calculus would play in the creation of a universe in which the dynamic movements of the planets could be explained mathematically. It was the discovery of calculus which would lead to the growth of classical mechanics and give birth to all branches of higher mathematics, to theoretical physics and chemistry, to celestial mechanics, to astrophysics, and to cosmology. Indeed, it is impossible for us to imagine how these ancient skygazers—who had little more than their own superstitions to guide them in trying to understand the heavens—were able to formulate even the most rudimentary of physical theories.

This is not to say that the ancient astronomers were fools because even the most primitive societies stared in wonderment at the whirling carousel spinning overhead. But they did not know how to explain the apparent retrograde motions of the planets or the periodic eclipses of the sun and moon except to attribute these apparent disappearances to the voracious eating habits of certain mystical monsters. But as time moved on and civilizations developed more sophisticated mathematics, they also became more adept at measuring the dimensions of the celestial sphere. Prior to the rise of the Greeks, however, the observational data was too scanty to permit astronomers to determine accurately the sizes of the planets or even hazard a guess as to the extent of the observable universe.

Even the Greeks were not immune to the tendency to place greater value on the creation of aesthetically pleasing principles than actually trying to explain the apparent chaos of the universe. Plato, for one, sought to geometrize the universe by supposing that the orbits of the planets corresponded with basic geometrical forms. He also offered a view of the heavens in which the sun was not much more significant than any of the planets. Aristotle did not question Plato's imaginative use of mathematics to map the cosmos even though he must have wondered about Plato's arbitrary arrangement of the solar system. Even so, very few of the Greeks were able to imagine a universe which extended beyond the bounds of the visible planets. Their perceptions of the cosmos were laughably modest, with most imagining the sun to be smaller than the earth and only a few planetary diameters away in space. But these inaccuracies were understandable given the absence of scientific instrumentation, the suffocating orthodoxy of Plato's astronomy, and the predominance of sheer flights of fancy as opposed to mathematical calculations as the basis for determining the magnitude of the cosmos.

It was only after the Greeks became acquainted with the accumulated observations of the Egyptians and the Babylonians that their own perceptions of the universe became more grounded in quantitative as opposed to purely speculative features. As the center of Greek civilization had shifted to Alexandria, Greek science and mathematics was profoundly affected by the growing belief that any theories about the universe should correspond with the observed motions of the heavenly bodies. To facilitate this effort, the Ptolemys, the rulers of Alexandria, built a great library known as the Museum which served as a forum for scholars to meet and study. The Museum also provided the funds for craftsmen to construct much more precise instruments for measuring the angular displacements of the planets and stars, which was an essential step for beginning to comprehend the true size of the universe.

The increasing use of mathematics in astronomy transformed that science from one beset by fantasy and ungrounded speculation to something approximating a quantitative science. Hipparchus of Rhodes, who flourished about 150 B.C., tried to make sense of the mass of Babylonian and Alexandrian astronomical data and, in doing so, discovered the precession of the equinoxes. As Hipparchus did not pen any best-selling novels or sidesplitting plays, however, his name might have disappeared from history altogether were it not for the Alexandrian astronomer, Claudius Ptolemy, who incorporated much of Hipparchus's work into his own treatise, *The Almagest*. This work offered a geocentric vision of the universe which utilized a new type of mathematics, called trigonometry, to determine the distances and sizes of celestial bodies such as the sun. Indeed, Ptolemy's *Almagest* was so much more sophisticated than any other astronomical work that it would serve as the basic astronomical text until the time of Copernicus. The fact that this work continued to exert such a significant influence on astronomers for nearly 1500 years has been regarded as a prime example of a theory outliving its usefulness because Ptolemy's geocentric theory discouraged his successors from considering competing sun-centered models of the solar system. But this was more a failing of Ptolemy's successors than of *The Almagest* itself because it was a truly remarkable work due both to its detail and its mathematical foundations.

One of the earliest applications of trigonometry involved the calculation of the moon's distance from the earth. This work was similar in principle to Eratosthenes's calculation of the circumference of the earth. Eratosthenes believed that the earth was round and thus consisted of a surface having 360 degrees of arc. He also knew that the distance between the cities of Alexandria and Syene was about 500 miles and that the sun shone directly overhead in Syene during the summer solstice. In Alexandria, by contrast, the sun did not shine directly overhead at the same time but instead was displaced by about 7½ degrees. Eratosthenes correctly supposed that if he divided the 360 degrees which make up a circle by 7½ degrees, he would obtain a value (48) which, when multiplied by the 500 miles separating the two cities, would give the circumference of the earth. This figure, equal to 24,000 miles, proved to be remarkably accurate and, indeed, was not improved upon for nearly two millennia. Unfortunately for Eratosthenes, this discovery did not bring him riches or widespread fame or even his own talk show. But his name will forever be linked to that select group of early explorers who used mathematics to derive factual data about the universe. More surprising is the fact that more modern thinkers, particularly those in the medieval era, ignored the sheer brilliance of Eratosthenes's work and decided that the world was flat.

Eratosthenes was one of the first thinkers whose work hinted at the promise of trigonometry even though he did not offer a formal set of mathematical principles, this task being left to Hipparchus and Ptolemy. But there were other Greeks who were able to use angles to aid in the solving of practical problems, such as Heron who tackled the problem of how to send two crews of laborers tunneling through a mountain from opposite sides so that they would meet each other in the middle. When we think of how difficult it can be to find our way in the kitchen late at night with the lights off and maybe a hazardous toy or two scattered about the floor, we can appreciate the difficulty faced by workers in dark, dusty tunnels who were trying to link up with their counterparts approaching from the other side of the mountain. This endeavor was especially important because digging tunnels was very dangerous work and there was always a risk that one's laborers would die before finishing the job—which invariably complicated the relations between management and labor. But as with Eratosthenes's calculation of the earth's circumference, the solution for minimizing the amount of time needed to dig a tunnel was really quite simple. Heron imagined that the two crews were beginning their excavation on opposite sides of the mountain, which could be described using two points A and B. Heron then imagined a third point C which was located halfway between points A and B on another side of the mountain but which was in direct sight of the first two points. In other words, a laborer standing at either point A or point B would be able to see a laborer standing at point C because point C was far enough away from the surface of the mountain to be visible by both A and B at the same time, as shown in Figure 4. The angle BCA (or ACB) was a right angle (90 degrees) and Heron realized that the ratio of the two sides BC/AC was equal to tangent A, which is the ratio of the length of the side of the triangle opposite angle A to the length of the side of the triangle adjacent to angle A.

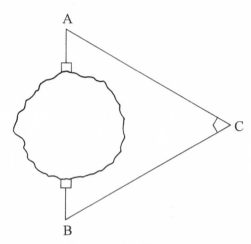

Figure 4. Two crews working from opposite sides of a mountain use a common point C (which forms a third angle of an imaginary triangle) and are able to use the basic trigonometric principles offered by Hipparchus so that they meet in the middle of the mountain.

Heron could then look up the trigonometric tables created by Hipparchus and Ptolemy to determine which angle A had the tangent value equal to this ratio. Once this value was determined, the laborers would then tunnel along a path oriented in that direction and would, if all went well and they did not drown in a subterranean river, meet somewhere in the middle of the mountain.

One can easily appreciate the benefit of being able to dig a straight tunnel as one does not have to allow a crew to meander through the bowels of a mountain trying in vain to find the counterpart crew. After all, time is money and, in the case of the primitive working conditions under which these crews labored, lives saved. Even today, engineers still use some of the same mathematical principles as those offered by Heron to dig tunnels. Of course today's engineers have to deal with uppity union laborers and do not have the benefit of being able to bully their slave workers but such are the drawbacks of living in today's modern society. Fortunately, the development of blasting and mechanized excavators has minimized the need to beat the workers senseless to motivate them to dig faster.

Hipparchus and Ptolemy did not begin their work in trigonometry with the idea that they could provide posterity with a better way to dig tunnels or draw triangles to amuse children at birthday parties. Instead, they had much more important things on their minds—measuring the dimensions and distances of the celestial bodies. This was not a simple task to undertake because Hipparchus and Ptolemy had to invent their own methodology and their own mathematics even though they did have the benefit of Eratosthenes's work in calculating the earth's circumference.

Knowing the earth's circumference was important because it made it possible to determine the diameter of the earth and, hence, the radius from its center to its surface (about 4000 miles). To take a stab at calculating the distance from the earth to the moon, one would essentially have to use a more sophisticated version of Eratosthenes's method. If we were to jump into the fray, we would begin by imagining that there was a point P on the surface of the earth. If we stood on point P (after pushing aside a malcontent who refused to get out of the way), we would look straight up and see the moon directly overhead in space. We could thus draw a straight line from point P up to the moon which we would designate as lying at point M. We would further imagine that this line would begin at point M, extend downward through point P, and onward to the center of the earth where the line would end at point E. Still standing at point P, we would put on our hiking boots and whistle a merry tune and walk to another point P^* which would be the furthest possible point at which we could still see the moon while it was directly over point M. Of course this example assumes that the moon is not moving at all while we are hurrying over from point P to P^*. To do so we would have to be able to walk an incredibly brisk pace of 6000 miles per minute or so which is not a realistic pace for most people. But this is a hypothetical example designed to illustrate basic trigonometric principles and as such is not overly concerned with tracking the speeds of world-class race walkers. In any event, we would now find ourselves standing at point P^* watching the moon hovering directly over point P and almost being able to peer into our neighbor's bathroom through a sheer curtain in a nearby apartment building. We would now draw a second line from point P^* up to point M, thus forming

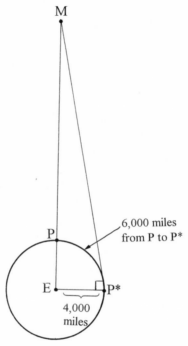

Figure 5. The trigonometric model used to calculate the distance from the earth to the moon.

two intersecting lines at *M*. We would then imagine a third line extending from point *E* at the center of the earth to our point *P**, thus creating the triangle *EMP** as shown in Figure 5, which could also serve as a snappy monogram on one's dress shirts even if one's first name was Charles or Sammy.

Thanks to the fine work of Eratosthenes, we can begin our attempt to calculate the distance to the moon—the length of side *EM* of our triangle—by finding the distance for the side *EP**, the distance from the center of the earth to its surface. This is where our knowledge of cool formulas such as the circumference of a circle ($2\pi r$) comes in handy because we can set it equal to Eratosthenes's figure for the circumference of the earth of 24,000 miles. We thus find that *r* is equal to about 4000 miles, which is the approximate length of EP^*. Now we know the length of the smallest side (EP^*) of the triangle, but the two remaining sides remain unknown.

If we were to slice the earth open and trace the paths of the intersecting lines EMP^*, we would see that they touch the surface of the earth at P^* and P. Connecting them is an arc whose endpoints are P^* and P; this arc is merely the geometrical representation of the surface of the earth that we just traversed in our snappy hiking boots. If we know that the length of the distance between P^* and P is 6000 miles (due to our having thrown down bread crumbs at mile intervals during our trek), then we would know from Eratosthenes's work that we had walked one-quarter of the earth's circumference since Eratosthenes had previously calculated it to be 6000 miles. Not wanting to concern ourselves with all of the members of our party who had dropped dead from exhaustion during our journey, we would sharpen our pencils and begin "crunching numbers." Actually, there are not too many numbers to crunch once we know that we can determine the angle E once we have measured the distance (arc) from P^* to P. In his book *Mathematics and the Physical World*, the mathematician Morris Kline points out that this angle E should be considered as the difference in longitudes between P^* and P so that knowledge of the longitudes of these two points will provide us with the angle E. By using Hipparchus's system for locating points on the surface of the earth using longitudes and latitudes, one can determine the angle E, which in Kline's example was 89 degrees 4'12" or just 55'48" shy of a right angle. This angle in turn tells us that the angle E is just a pinch less than a 90-degree right angle. Because we are mathematical geniuses who know that the sum of the angles of any triangle in Euclidean space is equal to 180 degrees, we can see that the third angle of our triangle EMP^* is going to be very small since the other two angles MEP^* (89 degrees 4'12") and EP^*M (90 degrees) total nearly 180 degrees by themselves. We might find ourselves shocked at the paltry size of angle P^*ME which is less than a single degree but it illustrates a fundamental rule of using these principles to calculate the distances of celestial bodies: The smaller the distance of the angle M for any celestial body, the farther away it is from the earth.

At this point you may be wondering if we are actually going to calculate the distance of the moon from the earth. Well, we do not want to deprive you of the full value of your entertainment dollar so we will now engage in the practice of trigonometry. There are three important ratios in trigonometry which all relate to the ratios which can be formed in a given right triangle. Those who frequent telephone dating services will recognize the three familiar terms: sine, cosine, and tangent. For those who are more economical with their pen and paper, these three terms can be referred to in a sort of mathematical shorthand as the sin, cos, and tan of an angle.

Of course you want to know how one goes about determining the sine, cosine, and tangent of an angle because this is one of those things one must learn in one's lifetime in order to be considered a truly cultured individual. If we imagine that we have a right triangle with three sides which can be described by its points *ABC*, then we can determine the sine of angle *A*, for example, by dividing the length of the side opposite angle *A* by the hypotenuse of the triangle, which we would write as *BC/AB* as shown in Figure 6.

Now we can risk a migraine headache by moving on to the cosine of angle *A*, which merely involves dividing the side adjacent to angle *A* by the hypotenuse of the triangle, which, in this case, would be *AC/AB*. So the difference between the ratio which gives us the sine for angle *A* as opposed to the cosine for angle *A* is that we have merely swapped the line *AC* for the line *BC*; the hypotenuse in

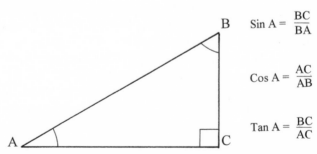

Figure 6. The ratios for determining the sine, cosine, and tangent of an angle.

both cases remains the same. Sounds great, right? Well, at the risk of bringing on a massive cerebral hemorrhage we would then proceed to the third and final of our three trigonometric ratios: the tangent of angle A. Here, we would divide the side opposite angle A by the side adjacent to angle A, which we would describe with the ratio BC/AC. The tangent of angle A would thus have the same numerator as the sine of angle A but a different denominator (AC instead of AB).

Now we could spend the next 50 pages talking about the historical evolution of these three terms and the origins of ancient Mesopotamian dialects in order to provide you with a truly eclectic background which can only be obtained in books that are far too important to be made into major motion pictures, but there is no need to cram too much knowledge into our heads as the brain can only absorb so much information before it explodes, rendering its owner capable of performing only the most rudimentary tasks such as network television programming and political office. Besides, trigonometry is deceptively simple and its practitioners take great pleasure in intimidating their new students by pointing out that the word "trigonometry" has five syllables, making it almost twice as hard as "algebra," which has only three syllables. But the worst of the work in trigonometry—the calculations of the sines, cosines, and tangents of various angles—has already been carried out by Hipparchus and Ptolemy, who apparently lacked lighter diversions such as television and charades. Indeed, it is a historical fact that neither Hipparchus nor Ptolemy had the time to engage in any games at all even though they would have undoubtedly enjoyed bridge or poker.

Needless to say, our discussion has taken us somewhat off the trail of our effort to determine the distance from the earth to the moon. At this point, we return to our trigonometric ratios and find that the cosine of angle E (89 degrees 4'12") is equal to 0.0163. By multiplying this quantity by the unknown quantity EM as shown in Figure 5 and setting it equal to the radius of the earth (4000 miles), we get the equation $0.0163\ EM = 4000$. By dividing 4000 by 0.0163, we arrive at a distance for EM which is equal to 245,000. Although this figure for the distance of the moon is not exact and has since been improved upon by our own more modern measuring techniques, it is remarkably accurate given the primitive astronomical and mathematical technologies which existed at the time and coupled with the reliance of these ancient astronomers on naked-eye observations.

This calculation of the distance from the earth to the moon was also helpful in forcing people to discard some of the more nonsensical notions they had about the universe and considering that it might consist largely of an ocean of empty space through which the planets and stars careened and twirled through eternity. Of course there was still much room for disagreement among those who believed that the earth was the center of the cosmos (and they themselves the center of the world) and those few renegades who thought that the sun was more deserving of that honor. But that was a battle which had to be left for another day when Copernicus would dare resurrect the heliocentric theory of Aristarchus and swing it like the stone in David's slingshot against the ignorance of the papal authorities. But as Copernicus would allow his theory to be published only in the last years of his life and, in fact, received a copy of his book on the same day he died (thereby putting a crimp in any plans he might have had for the royalty checks), it fell to others to carry out the battle to determine whether the sun or the earth was the center of the solar system, let alone the entire universe. The point of this digression is that once the initial calculations of the distance of the moon from the earth were made, humanity was forced to begin thinking that the heavens were far more spacious than had been previously imagined. The heavens had formerly been believed to be a kind of black ceiling peppered with pinpricks of lights which hung like a canopy over the surface of the earth. The realization that the moon was nearly 30 times farther away than the width of the earth's diameter itself torpedoed this notion of the earth as being a kind of stage around which the carousel of the heavens turned.

Hipparchus used the exact same method described above to calculate the distance to the sun. The problem Hipparchus encountered was that the sun's greater distance meant that he had to work with a value for E which was even closer to 90 degrees—more specifically, 89 degrees 59'51". This meant that the angle PMP^* shown in Figure 5 would be even smaller than with the moon. Because the technology of the time was crude and astronomers were forced to rely on their own eyes for their investigations, it was inevitable that their predictions of the distances of celestial objects other than the moon would be riddled with inaccuracies. Hipparchus, for example, underestimated the distance from the earth to the sun (which we know averages about 93,000,000 miles) by about 83,000,000 miles. While such inaccuracies might be welcomed by companies seeking to downplay corporate losses in their annual reports, they would not bring great cheer to our astronomers who, though inclined to round off big numbers to simplify their calculations, do not like to be so far off the mark. A more accurate determination of the distance to the sun naturally had to wait for the development of instruments which could measure more precisely these angles.

It is admittedly an exciting thing to know that you can, in a pinch, measure the distance from the earth to the sun so long as you have your trigonometric tables handy. But these early astronomers realized that they could go one step further by modifying the techniques for determining the distances to celestial bodies and therefore determine the diameters of these bodies themselves. How might someone like Hipparchus, who had lots of time on his hands, go about determining the diameter of the moon? In a sense, it would involve the same approach we used to determine the distances of the moon and the sun from the earth but we would be reversing the process. First, Hipparchus would draw a straight line from his position on earth (which we would denote as P) to the center of the moon (which we would denote as M) to create the line PM. After having taken a break to mark up his copy of Euclid's *Elements* and cube the squared sides of the formula for a right triangle to confuse future generations of mathematicians, Hipparchus would draw a second line from his position to a point on the edge of the moon (which we would designate as Q) to create the line PQ. The distance between Q and M would thus be equal to the radius of the moon because it would begin at the edge of the moon and end at its center as shown in Figure 7. We know from our previous discussion that the radius of the earth is 4000 miles and thus has a diameter of 8000 miles.

Although one could diagram this problem nicely with a can of black spray paint on the plush red carpet of the living room of a

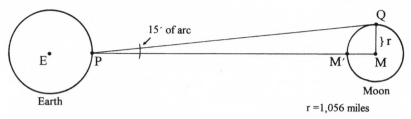

Figure 7. The trigonometric model used to calculate the diameter of the moon.

nearby mansion, the approach is fairly simple. Hipparchus would need to calculate the angle separating the two lines *PQ* and *PM*. Using his best estimate of the difference in degrees between a line passing straight through the center of the moon (*PM*) and a line barely grazing its edge (*PQ*), he might determine that the angular difference was surprisingly small—no more than 15′—which is about ¼ of 1 degree of arc or 1/1450th of an entire circle. This is an amount of arc that is barely perceptible to most persons but we are fortunate that Hipparchus, despite the demands of the social circuit of Rhodes (which at the time he lived was a wealthy commercial center), did have both the eye and the mathematical skills to make a fairly accurate calculation.

Hipparchus was followed by other very clever persons who, after calculating the 15′ of arc separating the edge of the moon from its center when viewed from earth, retrieved the figure of 245,000 miles separating the center of the earth from the center of the moon and lopped off 4000 miles to begin it at the surface of the earth. Having determined that the radius of the moon (which we denote with the line *QM*) was perpendicular to the line *PQ* extending from the surface of the earth to the edge of the moon, the next step was to determine the sine of 15′ of arc. What is the sine of this angle? Well, as we learned earlier, it has nothing to do with horoscopes or traffic engineering. Instead the sine of an angle (in this case the angle of 15′) is, as we saw above, equal to the ratio of the line *QM* (which is our unknown quantity) to the line *PM* (which we have determined to be equal to 240,000). By manipulating the equation, we obtain the equation $QM = 240,000 \sin 15'$. As Hipparchus was kind enough to calculate the sine of 15' as 0.0044, we multiply 240,000 by 0.0044 which gives us 1056 miles—which is very close to the generally accepted value of the moon's radius. Doubling this figure would give the diameter of the moon as 2112 miles.

These principles, which are so neatly and clearly summarized in Kline's book, were of tremendous help to the astronomers who followed Ptolemy. However, this method was clearly limited in its usefulness to those nearby bodies in the solar system for which a discernible angle could be measured. It also had the additional limitation that one could only calculate the distances from the earth to the moon or the earth to the sun or the earth to one of the other nearby planets. What if one wanted to calculate the distance between the moon and one of the other planets or the sun? Certainly it would be a handy thing to know the distances among the various planets of the solar system if one were to build a spaceship and begin an interplanetary ferry service. But it would also have important consequences for our theories of the universe because it would enable us to determine the distances between various celestial bodies at different points in time and thus, through trial and error, devise a model of the solar system.

The obvious difficulty with carrying out such calculations is that the measurements of the angles becomes a question of greater precision, with the vertices of the triangle being the three celestial bodies themselves (e.g., earth, sun, Mercury, or earth, sun, Venus). Certainly the determination of these values was beyond the capabilities of the early astronomers because they lacked accurate measuring devices and modern telescopes. Nevertheless, their effort to map the cosmos, given the prevalence of superstition and ignorance which dogged the efforts of all persons who sought to understand the universe as something more than merely the arbitrary and capricious actions of a pantheon of clay-footed gods, was a remarkable achievement. There were very few colleagues with whom one could share one's discoveries or draw moral support or even hide out with when the local mobs worked themselves up into an anti-intellectual rage and went on a rampage with torches and pitchforks. So the process of discovery and conjecture, flawed though it might have been at the time in view of the efforts by various individuals to verify and bolster Ptolemy's geocentric (earth-centered) theory of the universe, was a solitary one. The discoverer was forced to follow his own convictions as to the rightness of his views and cast aside the stifling prejudices of the day which would have deterred many persons who were less sure of the validity of their cause or who lacked the desire to understand the workings of the cosmos.

Once Hipparchus and Ptolemy began making measurements of the distances from the earth to other celestial bodies, their findings, though marred in some cases by significant errors—such as Hipparchus's underestimation of the distance from the earth to the sun by a factor of nine—caused the comfortable, compact vision of a cosmos of a few nearby planets and a burning orb revolving around the earth to begin to crumble. Because the apparent size of the sun, when viewed from the earth, is fairly small, it was difficult for people to accept the idea that the sun itself might be bigger than the earth—let alone have a volume which could contain a million earth-sized planets. But the power of trigonometry compelled scientists such as Hipparchus and Ptolemy to conclude that the earth was not the largest celestial body in the universe. However, they could not bring themselves to consider the other logical consequence of this revelation—that the earth might not be the center of the universe. After all, it was not the most logical of conclusions to place a comparatively small planet at the center of the universe. But it was admittedly not logic that prompted the development of the geocentric cosmology in the first place but instead the belief that the earth occupied a unique position because it was the home to humanity. Because it would be more than 15 centuries before Isaac Newton would be hit on the head with an apple and develop both a headache and his theory of gravitation, there were very few persons who seriously questioned the geocentric theory. Had they known that all bodies are gravitationally attracted to each other, they would have seen how unrealistic it was to imagine comparatively massive bodies continuing to orbit around the planet earth. In time, they would have been driven to the opposite conclusion that the center of the solar system would have to be the most massive body—the sun.

The limitations of the astronomical technologies of that time precluded Hipparchus and Ptolemy from making accurate measurements of the distances of the other planets, such as Mars and Jupiter. The angular measurements became smaller as one moved outward from the earth to more distant celestial bodies. Hipparchus and Ptolemy did not know about the outer planets of the solar system, such as Uranus, Neptune, and Pluto, but the angles of measurement for those bodies would have been too small for them to detect. If Hipparchus and Ptolemy had possessed even grander ambitions and sought to measure the distances to the stars, they would have found their task to be impossible. Whereas the distances of the planets close to the earth can be measured in terms of a few light-minutes, the nearest star (other than our sun) is more than 4 light-years away. Trying to use the naked eye to measure the angles of these stars would have been about as easy as learning high-energy particle physics through a correspondence course.

The efforts to measure more distant celestial bodies did not advance appreciably in the ensuing centuries. The Roman legions were too busy wreaking havoc throughout the Mediterranean world to pay much attention to measuring astronomical distances. Similarly, the astronomers of the medieval era were more concerned with fitting their observations into the existing religious orthodoxy than they were with modeling the universe. Indeed, it was only after the 18th-century English astronomer James Bradley modified some of the basic trigonometric triangulation techniques invented by the Greeks to calculate the velocity of light that astronomers were able to begin making accurate calculations of the distances to the stars.

For those brave persons who would like to go out and calculate the velocity of light using trigonometry, Bradley's use of mathematics should prove to be particularly instructive. He imagined that the earth was at one end of the hypotenuse of a right triangle and the star (the light source to be measured) at the other end of that very same side. The third vertex of the triangle was a point at which the earth would cross paths with the beam of light to be measured. Bradley determined the angle at which the earth was initially located in this imaginary triangle to be just a few seconds less than 90 degrees of arc and the value of the tangent of this angle to be about 10,000. He then calculated the velocity of the earth to be about 18.5 miles per second based on the known distance of the sun from the earth and the formula for the circumference of a circle (Bradley had assumed the orbit of the earth to be circular instead of elliptical in order to simplify his calculations; this assumption did not greatly alter the accuracy of his calculations). By multiplying the known velocity of the earth by the tangent of the angle, Bradley calculated the velocity of light to be more than 180,000 miles per second and thus provided astronomers with an indispensable tool for measuring interstellar distances. Of course the English public would have been much more impressed had Bradley invented high heels or dental hygiene but such are the vagaries of fame and fortune.

Opening the Heavens

The work of Hipparchus and Ptolemy forced humanity to reconsider its perceptions of the universe by showing that the distances between the earth, its moon, and the other nearby bodies of the solar system were much greater than had been previously imagined by the early astronomers. But this success prompted the Greeks to begin thinking about even greater intellectual feats such as devising an actual geometrical model of the universe. We know that the Greeks had always been enamored with the immutable principles of Euclidean geometry so it was probably only a matter of time and the consumption of the right vintage of wine before some of Greece's greatest thinkers were prompted to try to superimpose the basic principles of this geometry upon the physical universe. Plato, perhaps the best known of the Greek philosophers, was convinced that the spectacle of the heavens could be explained in terms of basic geometrical principles. In a way, Plato was remarkably prescient because Einstein's theory of general relativity did propose that the structure of the universe, and ultimately its fate, was based in part on the gravitational curvature of space. But Einstein's theory required the application of the geometries of Lobachevsky and Riemann and was not merely a fanciful speculation about the relationship between Euclidean geometry and the structure of the universe. Plato's work, on the other hand, was more of a random application of geometrical principles to the universe which could not be verified in any meaningful or acceptable way. Not surprisingly, Plato's efforts to "geometrize" the cosmos were not wildly successful and, indeed, ran aground because his proposals did not really lead to any meaningful conclusions. It was only through the efforts of Hipparchus and Ptolemy that any real progress was made in developing a model of the cosmos which could explain the apparent motions of the planets.

But any impetus given to the progress of astronomy in Greece and, indeed, the Eurasian land mass as a whole, dissipated with the collapse of the Roman Empire in 476 B.C. Although the achievements of the Greeks had not really depended on the central authority provided by Rome, the continued evolution of astronomy and mathematics became much more problematic with the end of the relative peace provided by the Empire's soldiers. After all, it is much easier to keep the peace when you have a lot of burly louts who can club any troublemakers senseless at a moment's notice in order to preserve domestic tranquility.

Because the Romans did not really add a great deal to the fields of astronomy or mathematics (even though they did make contributions in many other areas including architecture, public works, and elaborate orgies), few mathematicians and astronomers felt a strong pang of loss when the Roman legions ceased to roam across Europe. However, the disappearance of the Romans was followed by the emergence of the barbarian tribes. This was not particularly good news for those people who like their next-door neighbors to be neatly groomed, articulate individuals who settle their disputes with calm reasoning instead of the swing of a sword. Sadly for those persons who value good table manners and the plentiful use of "pleases" and "thank-yous," most barbarians were coarse, swarthy, nasty, violent brutes who enjoyed sacking cities. Very few were schooled in the finer points of etiquette and even fewer would have had a clue as to which fingers should be used to lift a cup of English tea up to one's lips. Not surprisingly, these barbarians were not the sort of plodding, methodical types which every society needs to build its buildings and infrastructure; they much preferred the gay life of racing at breakneck speed on horseback and torching villages and beheading unarmed peasants. As a result, they were not the best of candidates for creating a new post-Roman Empire society. But they were the perfect persons for ushering in that period of history known as the Middle or "Dark" Ages in which Europe was plunged into a mind-numbing monotony of subsistence agriculture and fragmented politics.

Some would liken the barbarians to today's revenue agents because the barbarians were not really interested in the sciences or the arts. They cared little for music or philosophy and would have been more comfortable impaling Plato on a gate instead of reading his treatises on politics or ethics. But even barbarians are not islands unto themselves and they, despite their anti-intellectual predispositions, were gradually "softened" by the civilizing influences of the Christian Church, which remained the bulwark of Western civilization during this era with its monasteries and churches serving as the repositories of numerous works of literature and art. Many of the barbarians were also converted to Christianity which, over several centuries, had an extraordinary pacifying effect on these once unrepentant plunderers and looters.

One reason that the work of Ptolemy endured for 15 centuries was that there was little interest throughout the medieval era in considering alternative astronomical theories or even, for that matter, studying the existing theory. Although advances in mathematics were made during this time by the Arabs and, to a lesser extent, the Chinese, very few persons devoted their energies to the deeper study of mathematics. So the legacies of the past endured due to the lack of any real interest in augmenting, let alone challenging, the prevailing perceptions of the universe. This would not have been a happy time for textbook publishers (had books existed at the time) because there would have been no reason to issue new editions in science and mathematics texts for centuries at a time.

The Christian Church assumed the dominant political role in medieval Europe, with the papacy wielding greater power than any single monarch. But with this power came a stifling of free and independent thought in Europe in which a literal interpretation of the Bible was seen as the final arbiter of all questions—including those relating to issues which we would nowadays consign to science and mathematics. This was a handy approach for those persons who liked the idea of not having to collect an entire library of books in order to claim a well-rounded knowledge of the world. But it was a disaster for the progress of science because it encouraged the dogmatists of the Church to cast aside all things Greek as being profane. It also prompted the Church scholars to adopt some ludicrous theories about the world and the universe as a whole in order to conform to the metaphorical accounts of the Bible. Perhaps the most telling example of the mauling of Greek science was the emergence of the idea of a flat earth, which gladdened the hearts of those who feared that they would otherwise fall off the underbelly of a spherical planet. Many of the Greeks had written extensively on the spherical nature of the earth and more than a few Greeks had concluded that the curvature of the earth was responsible for the recession of the shoreline observed from ships sailing out to sea. The papacy's brain trust, by contrast, was convinced that the earth must be flat because the now unfashionable Greek theorists had concluded that it must be round. They also took comfort in certain Biblical references which seemed to favor the idea of a flat planet. The wonderful thing about dogma is that one does not necessarily have to worry about satisfying a burden of proof; the Church theologians heartily subscribed to this point of view and pointedly ignored the works of the Greek astronomers and mathematicians which showed quite convincingly, as we have already seen, that the earth was round.

Fortunately for those persons who like their science to have some basis in observation and mathematics, much of the Greek legacy had been preserved elsewhere. Arab conquerors had swept through Northern Africa into Spain and Italy during the early medieval era, before the orthodoxy of the Christian Church had assumed such a powerful sway over the European continent, luckily preventing the complete destruction of the classic works of antiquity. Indeed, some of the Arab scientists had made their own important advances in mathematics and astronomy before these works were reintroduced into Europe. Not surprisingly, the papacy was not thrilled to see that many of Europe's thinkers were interested in the allegedly blasphemous works of Plato, Archimedes, Aristotle, and the rest of the Ionian gang. But after having been subjected to an unrelenting dose of theology and doctrine and having had their own inquiries about the universe or God or life itself ignored or even used as ammunition for charges of heresy, many of Europe's intellects were ready to become reacquainted with its classical legacy. Of course they were careful not to go running down the streets praising Hipparchus's ingenuity in calculating the diameter of the earth or even dropping a postcard to the Pope to ask him his opinion of the heliocentric theory proposed by Aristarchus. The reason for this circumspection was that the Church did not welcome such free thinkers with open arms; instead, Church officials preferred to penalize such individuals for daring to diverge from the officially sanctioned doctrines. Excommunication and burning were among two of the more popular choices, the latter proving to be a popular form of family entertainment in the Campo dei Fiori in Rome. Indeed, this was the fate that befell Giordano Bruno, an Italian mystic, who believed that the universe contained an infinity of worlds. The Church officials were not quite so enamored with Bruno's generous vision because it posited a cosmos in which there was no central place. As a result, no single planet's citizens could claim a favored status or to have been visited by Jesus Christ. So Bruno's vision, though not offered as a direct challenge to the Christian Church, did in fact pose staggering philosophical issues for the Church which could not be ignored. In a universe having an infinity of worlds, there seemed to be nothing to set the earth apart from the other celestial bodies. Bruno's concept also treated the idea of an earth-centered cosmos as an absurdity because one can have no center point in a universe extending indefinitely far in all directions. So the Church, when faced with these challenging ideas, did the only rational thing it could do and hauled that troublemaker Bruno out to the square and set him on fire. Some of us might pine for such simple times in which these handy remedies were available for dealing with malcontents but ours is a world in which we have civil rights laws that unfortunately get in the way of the swift and entertaining dispensation of justice.

Bruno's death occurred at a time when the power of the Church was beginning to wane. Its ideologues were desperately trying to hold on to the world of the past in which their own often scientifically vacuous ideas went unchallenged. This was also a time in which a young Polish student named Nicolaus Copernicus had completed his studies in mathematics and science at the University of Cracow and had recently enrolled at the University of Bologna where he began studying many of the Greek classics which had been forbidden by the papal authorities. He also learned how to make astronomical observations and, after studying law and medicine, accepted a clerical position at the Frauenberg church in East Prussia. By this time, Copernicus had learned a great many things about a variety of subjects. He became a mediocre cleric and a very competent naked-eye astronomer. During the ensuing years, Copernicus spent many evenings studying the skies and trying to determine if there was not a simpler and more elegant model of the cosmos than the cumbersome geocentric model that had been offered by Hipparchus and Ptolemy.

Much of Copernicus's dissatisfaction with the geocentric model stemmed from its sheer complexity. Originally, Hipparchus and Ptolemy had started with a very simple and sensible idea: They lived on the planet earth and, therefore, the planet earth was located at the center of the universe. This was perhaps not the most scientific approach to take in constructing a cosmological theory but it certainly cut to the heart of the matter and did not waste time gathering data and testing hypotheses. Ptolemy had crafted a very elaborate theory which attempted to account for the apparent motions of the celestial bodies in an earth-centered universe. It posited that each of the known planets moves in circles called epicycles which revolve around the sun; the sun in turn moves on a circular path called the deferent which revolves around the earth. But the need to make the theory correspond with the observed motions of celestial bodies in the sky necessitated that a very complicated array of epicycles be constructed, which seemed to be extraordinarily clumsy to Copernicus. Copernicus's frustration with this model made him search for alternatives and, at some point, he returned to Aristarchus's long-dormant theory which had posited a sun-centered universe. Copernicus found that if he assumed the planets move in perfect circles around the sun, he could construct a model of the solar system which not only corresponded with his observations but also did not require the epicycles that encumbered Ptolemy's model. This is not to say that Copernicus's model could explain all of the apparent move-

Figure 8. Nicolaus Copernicus. Courtesy of AIP Emilio Segre Visual Archives.

ments of the celestial bodies, but it represented a significant improvement over Ptolemy's system, both in terms of predictive accuracy and simplicity.

But even in the mid-16th century (c. 1530) when Copernicus first published a pamphlet highlighting his heliocentric theory, he was well aware that the papacy had no aversion to roasting those persons who would dare challenge the fundamental tenets of Christian dogma. Certainly Copernicus's theory represented a poke in the eye of the Church, which continued to cling to Ptolemy's geocentric universe with all the tenacity of an Internal Revenue Service auditor. But somewhat curiously, Copernicus was encouraged to publish his work by Cardinal Nicolaus von Schonberg, the Archbishop of Capua, so that it could be made available to a wider audience. Several other friends of Copernicus, most notably George Rheticus, who was a professor at the University of Wittenberg, oversaw the publication of Copernicus's great work, *De Revolutionibus*, which, as luck would have it, was handed to Copernicus as he lay on his deathbed.

With Copernicus already dead, there was little that the Church would have gained by burning him so it fell to the dogmatists to try to stifle the spread of news of his work. But the genie was already out of the bottle and Copernicus's ideas were eagerly embraced by those who were fortunate enough to obtain a copy of his book. The widespread acceptance of his ideas, however, was slowed to some extent by the errors in his theory. The most glaring problem was that its predictions did not correspond exactly with the observed orbits of the planets—even though it represented a significant improvement over the system of Ptolemy. Indeed, Copernicus had been so frustrated with the inability of his theory to predict accurately the complete orbits of the planets that he had adopted some of the very same gimmicks used by Ptolemy—most notably, the epicycles and deferents—to try to improve the accuracy of his model. These attempts to tinker with his model did not really solve his dilemma but merely underscored the incompleteness of his heliocentric theory. So Copernicus left this earth having offered a model which was fundamentally correct in placing the sun at the center of the solar system but which still had problems beyond his capabilities to resolve.

The problem lay in Copernicus's insistence that the planets travel in circular orbits around the sun. His insistence that the planetary orbits are circular was due to both the lingering influences of the Greek geometers such as Plato, who sought to model the universe on a succession of geometric shapes, and the idea that the planets are part of the sacred heavens and thus must move in perfect circles around the sun. Despite Copernicus's polyglot educational background, it seems that his religious training had been intense enough to make him subjugate his instincts as a scientist (that circular orbits might not be the answer) in favor of his theological leanings (that circular orbits are the only appropriate ones because they are perfectly round).

It fell to the astronomer and mystic Johannes Kepler, whose career began nearly 50 years after Copernicus's death, to put the heliocentric theory on a sound footing. He made a thorough comparison of the theory's predictions and the astronomical data which had been compiled by the Dane Tycho Brahe during the last quarter of the 16th century. Kepler could not have selected a better compiler than Brahe, who was perhaps the greatest naked-eye astronomer of all time. Kepler had worked as Brahe's assistant and, as luck would have it, happened to be waiting in the wings when Brahe dropped dead in 1601. Indeed, Kepler had so impressed Brahe's employer, Emperor Rudolph II, with his abilities that he was appointed by the Emperor to be the new "Imperial Mathematician" and assumed Brahe's position. Kepler soon learned, much to his dismay, that his most important duties had little to do with astronomy or mathematics; he spent most of his time divining the horoscopes of the Emperor and his courtesans. But Kepler was not one to stand in the way of the Emperor's happiness even though he believed that the idea of predicting one's fortune based on the random arrangements of stars was absurd.

Fortunately, Kepler lived at a time when emperors were able to have their imperial mathematicians spend much of their days engaged in heavy-duty horoscoping without having to worry about intrusive exposés or detailed audits. After all, one could jail journalists and auditors in those days without attracting very much attention. The demands on Kepler's time were not overly burdensome; he was able to enjoy ample leisure time in which he could study Brahe's tables and thus feel he was actually carrying out the duties befitting an imperial mathematician.

But Kepler did not have very much time to enjoy his newfound membership in the rarefied air of Emperor Rudolph's court because the government was bankrupt. Moreover, Emperor Rudolph, though he greatly appreciated Kepler's services as an astrologer, was not overly concerned with paying Kepler. Though Kepler did not live a lavish lifestyle, he had to scrape and scrimp to eke out a living even though he spent much of his time predicting the fortunes of the members of the court at the royal residence. His predicament was made more difficult when his wife suffered a nervous breakdown and went completely mad. Although she lingered on for a while, she eventually died, followed in death by one of the couple's children. Political upheavals compounded Kepler's problems and made the collection of his salary even more of an adventure. Even though Kepler was a first-rate horoscoper, he found fewer and fewer paying customers. Yet it was astrology, despite the later fame he would realize as an astronomer, which enabled Kepler to survive during his waning years in Rudolph's court. Kepler eventually reached the point where he could no longer serve as a charitable benefactor to Emperor Rudolph. In 1612 he accepted a teaching position at the University of Linz. But ill fortune continued to plague him. He remarried but found marital bliss to be elusive. Two more children died and Kepler's new wife made it her mission in life to complain constantly about Kepler's inability to earn a satisfactory living. Kepler, for his part, perhaps seeking solace from the friendly advice which was being offered by his wife on a daily basis, immersed himself in his work. But his difficulties were heightened by the hostility of the local citizenry, who were suspicious of Kepler's Lutheran background. Eight years after Kepler came to Linz, the city fell to the invading army of Duke Maximilian of Bavaria, who was himself a Catholic. Not wanting to see Kepler bored at the prospect of not being persecuted, the Duke's officers stepped up to the plate and did their best to make his life even more unpleasant. All of these difficulties weighed Kepler down physically and emotionally, precipitating a steady but inexorable decline in his health. For the remaining 10 years of his life, Kepler would struggle to keep his family together and carry out his work.

Kepler was convinced that all the workings of nature could be explained by basic mathematical principles, even though he still dabbled in astrology and mysticism. But he had enough confidence in this conclusion to abandon the gimmicks of earlier cosmologies such as the epicycles and deferents which had infested both Ptolemy's geocentric system and Copernicus's heliocentric model. Despite this conviction, however, he was not quite sure how he should proceed in devising a more appropriate model of the cosmos. Having access to Brahe's astronomical data as well as a body of his own observations, Kepler was much more of a roll-up-your-sleeves astronomer than most of his predecessors, particularly the Greek philosophers such as Plato, who were content to devise geometrically based theories of the world without any real concern as to whether there was any evidence to support these theories. Kepler, on the other hand, favored observations and factual data to lofty but unsubstantiated theorizing and thus proceeded to work in the opposite way. He studied the data for many years and then tried to develop a model of the solar system that was in agreement with that data.

Kepler discarded the circle as the orbital path of choice in astronomy because he could not reconcile the observed movements of the planets with the circular orbits predicted by Copernicus's model. Based on his studies of the orbit of the planet Mars, Kepler became convinced that the ellipse—not the circle—was the correct shape. Dropping the circle was a revolutionary break with tradition as it freed astronomy from the geometric formalism which had plagued it for nearly 2000 years. The adoption of the ellipse formed the basis of Kepler's first law of planetary motion, which states that the planets travel in ellipses around the sun.

Feeling that one law of planetary motion was not enough to ensure true immortality, Kepler established several other physical laws which would bear his name. He soon discovered that a planet moving around the sun does not move at constant velocity but instead moves faster as it approaches the sun and then slows down as it moves away from the sun. Applying this finding to the ellipse, Kepler concluded that the area covered in a given amount of time (if one were to draw an imaginary line from the moving planet to the sun) is constant, even though the velocity of the planet varies with its relative proximity to the sun. This means that the line would sweep out over equal areas of space in equal amounts of time. When the planet was near the sun, it would traverse a greater slice of the orbital arc. Conversely, when the planet was farther away from the sun, it would traverse a smaller slice of the orbital arc. This constancy in area covered per unit of time provided even more evidence for Kepler's belief that mathematics can describe the fundamental relationships of the celestial bodies. It also convinced Kepler that he might be able to find at least one more law of planetary motion to bear his name if he searched long enough.

Kepler's third law of planetary motion was the culmination of his search for an underlying harmony in the universe. He eventually derived a formula which would make it possible to calculate a planet's distance from the sun if one knew the period of time needed for it to revolve around the sun. This relationship was expressed mathematically by Kepler as T^2/D^3 where T is the time needed for the planet to complete one revolution around the sun and D is the mean distance of the planet from the sun. Because Kepler was thoroughly familiar with the mystical nature of certain numbers and convinced that the basic features of the universe could be described using mathematical laws, he viewed his third law as the Holy Grail of astronomy, describing, as it did, one of the most fundamental physical truths about the movements of celestial bodies.

The world did not give Kepler a standing ovation. Kepler still lived in a time when Europe was divided by religious and ethnic conflicts. The chaos of that era merely encouraged the Church to take a more unyielding stand against those who would dare to challenge its authority. Any hopes that Kepler might have had about being welcomed by those troublesome Protestants were quickly dispelled as they considered his work, like that of Copernicus, to be heretical. After all, the Protestants had split off from the Catholic Church because they viewed Rome as having strayed from a literal interpretation of the Scriptures. As a result, the Protestants would have extended a much warmer welcome to Kepler if he had devised three laws bearing his name which propped up the geocentric theory offered by Ptolemy. But one cannot always have everything one's way and the Protestants, try as they might, could not ignore the mathematical elegance of Kepler's works. Yet Kepler's views won only grudging acceptance; very few souls dared to publicize their support of his work or the Copernican theory as a whole.

With the hindsight of nearly four centuries, it is easy to look at the society in which Kepler and Copernicus worked and wonder how those who clung to Ptolemy's model of the universe could be so narrow-minded. After all, we are much more highly advanced as a species than our early 17th-century brethren as shown by our ability to demolish entire cities with a single bomb or to manufacture 12-blade pocket knives. But the acceptance of the heliocentric theory was not a simple matter because there were a number of objections which could be raised to it by scientists who did have solid credentials and respectable reputations. Foremost among these objections was the apparent lack of movement of the stars in the sky as the earth sailed through space. The geocentrists argued that the stars should appear to approach and then recede as the earth traveled around the sun. But the heliocentrists such as Copernicus responded that the lack of any movement by the stars was due merely to their vast distances from the earth. The farther away the object, the less it would appear to shift when viewed from the vantage point of a moving object such as the earth. The geocentrists were not terribly enamored with Copernicus's response because they did not believe that the stars were so far away that such movements could not be detected. Indeed, the prevailing view of the cosmos was one of suffocating compactness in which the earth was surrounded by a canopy of stars and planets overhead. The idea that these luminous objects might be located more than a hop, skip, and jump above the earth's surface struck many persons as ludicrous.

But even Copernicus had no idea of the true distances of the stars from the earth. Not in his wildest fantasies would he have supposed that the visible universe was as large as modern astronomers now know it to be. He would have found it almost impossible to conceive of the closest star to the earth being located about 4½ light-years away. Indeed, he would have had great difficulty understanding the concept of the light-year itself (the distance that light travels in a year; it has nothing to do with whether one restricts oneself to a low-fat diet for an entire calendar year) because there was nothing in the sciences or the religious dogma of that time which would have encouraged one to view the universe as an unimaginably vast ocean of space and matter. This was a view which would not have found great favor with many persons—whether inside or outside the Church—because everything was geocentric and introspectively-oriented. The intense religiosity of the era encouraged people to focus on their own internal salvation and to pay scant attention to the world around them. The obvious dangers of diverging from the Church's official line were also periodically made apparent with the persecutions of well-known individuals such as Galileo Galilei. Despite his own indomitable spirit, Galileo was placed under house arrest and forced to recant his views about the heliocentric system.

But Copernicus and Kepler would have been hard pressed to answer other objections to the idea that the earth spins around the sun. Why did the earth move at all? What actor or agent had caused it to move in the first place? Why did the people on a moving earth not see its atmosphere stripped away like the petals of a flower being buffeted by a gale force wind? Why did people themselves not fly off into space? What caused the earth to continue remaining in orbit around the sun?

The questions about the earth's movements through space and the forces that caused this movement in the first place were not easily answered and invited speculation. These questions ultimately led people back to the idea of a supernatural deity. The questions surrounding the continued adhesion of the earth's atmosphere (as well as its inhabitants) to the planet's surface could not be resolved using the Aristotelian physics which anchored the geocentric model. Finally, the fact that the earth remained in orbit around the sun necessitated the creation of an entirely new physics which would come in another century with the work of Isaac Newton.

But despite the objections raised by the geocentrists, Copernicus and Kepler pressed forward in their efforts to refine the heliocentric theory. They both were religious men who were convinced that the heliocentric theory was an even grander testament to the glory of God because it forced humanity to recognize that the universe was a far more extensive system of stars than had been previously thought. They did not concern themselves greatly with reconciling the literal interpretation of the Scriptures such as the uniqueness of Jesus Christ's time on earth with the heliocentric doctrine. For them, the divinity of the universe was reflected in the simplicity of the model developed by Copernicus and in the mathematical principles devised by Kepler. In essence, heliocentricity was viewed by Copernicus and Kepler as a further step in humanity's growing understanding of a magnificent universe which was only beginning to reveal some of its secrets.

Their approach was far more modern than that offered by any of their predecessors because Copernicus and Kepler were convinced that mathematics could be used to explain fundamental relationships in the cosmos such as the orbits of the planets around the sun. They believed that mathematics should conform to reality instead of reality being distorted to fit within the boundaries of the mathematics. We need only recall that Plato had become so enamored with the perfection of his geometry that he sought to "geometrize" the universe by mashing its observed manifestations, often to the point of absurdity, into the framework offered by his mathematics. He would not look at the movements of the planets and seek a mathematical relationship which could provide a universal expression for those movements. Instead he would seek to impose the pristine principles of Euclidean geometry onto the universe, proposing that the solar system consisted of a series of basic geometrical forms, one inside the other, which corresponded with the orbits of the planets and the sun around the earth. In a sense, Plato would have preferred to create the perfect, transcendental theory which did not explain perfectly the manifestations of the physical world. Plato viewed the physical world as a base, imperfect place. It was only in the intellectualized world of abstract thought that a truly "perfect" geometrized universe existed.

It fell to Galileo Galilei, the inventor of the telescope, to provide some observational evidence to bolster the heliocentric theory. Certainly it had already begun to attain some acceptance by astronomers (who were impressed with its simplicity) and sailors (who found it far easier to understand and use for navigation). But until Galileo turned his telescope to the evening sky, the acceptance of the theory was based as much on faith in its elegant simplicity and ease of use as it was on the astronomical data collected by naked-eye astronomers such as Tycho Brahe up to that time. But once Galileo began using his telescope, a curtain of darkness and ignorance was swept away and he was able to explore a universe in which some of the other planets such as Jupiter had moons revolving around them and the observed number of celestial bodies was far greater than anyone had previously anticipated. Even though a few gadflies such as Giordano Bruno had argued in favor of an infinite universe of worlds, Bruno's vision was derived from his own philosophical beliefs and was not based on physical observations. Once Galileo saw thousands upon thousands of stars which had never been seen by humanity before, the idea that the universe was far larger (perhaps even infinite) was considered much more seriously by astronomers and mathematicians. While Galileo's observations of everything from planetary moons to Saturn's rings to the bright band of starry light which we now know to be the Milky Way galaxy did not prove that Copernicus was correct, it did underscore the increasing tenuousness of the geocentric world view. The greater the number of planets and stars, the more unlikely it appeared that a single planet such as the earth could be the focal point of the entire universe or that our solar system could be regarded as unique or extraordinary.

The death of the geocentric theory did not occur overnight. The greater simplicity of the heliocentric doctrine and its ease of use coupled with its reliance on mathematics virtually compelled its acceptance by many scholars who might have objected to it on purely philosophical grounds. But the ultimate triumph of the heliocentric theory was not necessarily guaranteed as its proponents had to do battle with a medieval Europe riddled with superstitious beliefs such as witchcraft and demonology. Natural disasters were seen as examples of divine retribution instead of the outcomes of ongoing physical processes. The world was thought to be inhabited by witches who cast spells and engaged in cannibalism. The development of the Copernican theory encouraged the view of a world governed by physical laws—not the capricious actions of beings possessing magical and malevolent powers. That it ultimately prevailed was a testament to its own intellectual foundations because nearly every European court had a royal astrologer who predicted the fortunes of monarchs and nobility alike. Moreover, many of the leading thinkers of the day, including Roger Bacon and Kepler himself, seem to have had a genuine belief in the veracity of astrology. But the apparent sameness of physical laws throughout the universe as shown by the work of Copernicus, Kepler, and Galileo undermined and, indeed, destroyed the notion that the physical positions of certain stars relative to each other could affect the fortunes of humans. The discovery that there are many other star systems in the universe underscored the futility of pretending that a few arrangements of stars which together formed the zodiac could be used to answer all of the questions which might be posed by an individual. After all, there were clearly many other collections of stars which could also be added to the zodiac if one were so inclined. But it seemed rather absurd to suppose that every collection of stars should be a part of the zodiac and, undoubtedly, would have created a tremendous problem for the astrologers who had such a vested interest in a 12-sign zodiac. One could only imagine having to add two or three thousand more signs of the zodiac ranging from the sign of the walrus to the sign of the gnat and then trying to make some sense of the entire mess. Although it might encourage people to become more familiar with some of the more obscure members of the animal kingdom, it seems unlikely that it would have added a great deal to the utility or legitimacy of astrology itself.

The greatest contribution of heliocentricity may have been its reliance on the predictions offered by mathematical relationships and logical reasoning as opposed to those based purely on sensory perceptions. In a sense, Copernicus and Kepler asked us to refrain from accepting at face value the apparent images we perceive in the real world and instead see whether these knee-jerk beliefs can be understood and reconciled when placed beneath the lamp of mathematics and reason. This exhortation that we not necessarily trust our eyes without engaging in some measure of analysis is a signpost that continues to guide the sciences and shape the development of all mathematics—particularly the differential calculus.

CHAPTER 8

Differential Calculus

I should like to know if any man could have laughed if he had seen Sir Isaac Newton rolling in the mud.
—SIDNEY SMITH

Having wandered throughout the mathematical landscape and sampled arithmetic, algebra, geometry, and trigonometry without vomiting or even suffering from a bout of indigestion, it now remains for us to begin the exploration of the crown jewel of mathematics—the calculus. This is not to say that the calculus is the most difficult or the most complex of the various branches of mathematics but it does appear to be the most prestigious—at least in the minds of the lay audience. If you mention the word "algebra," most people will look as though they have just swallowed a frog. Geometry is usually regarded with greater fondness, perhaps because people enjoy the conceptual principles of geometry more than the mathematical manipulations which underlie algebra. This preference may relate to that old proverb that "a picture is worth a thousand words." People relate more easily to drawings of circles and rectangles and planes than to formalistic, symbol-laden equations that represent a greater degree of abstraction.

But like algebra and geometry, the word "calculus" has evoked its own images among those who have had the fortune to take years of mathematics classes and carry out tens of thousands of mind-numbing calculations to hone their quantitative skills. Calculus deals

185

with change and rates of change. It has a dynamic aspect to it that is missing from the comparatively static subjects of geometry and algebra. It is relevant to almost every modern technology and has proved to be of inestimable value to science and industry. It is as profoundly original as any branch of mathematics and, with the possible exception of Euclidean geometry, almost as well known.

What we call the calculus is the umbrella term under which the *differential* and the *integral* calculus are commonly grouped. Although we will consider each of these two distinct forms of the calculus in turn, it should suffice for now to discuss the reasons behind the creation of the calculus. Although its basic principles would be formulated independently by Newton and Leibniz in the 17th century, the need for a more sophisticated type of mathematics that could express rates of change had been recognized long before the proverbial apple fell on Newton's head. In particular, there was a growing need for mathematical tools which could express quantities undergoing constant change relative to each other such as the motions of the planets around the sun. In its briefest form, the calculus was devised to provide a concise answer to the question of how the changes in one variable (quantity) would affect the changes in a second variable (quantity).

This concept of rate of change should not be confused with the concept of change itself. How are they different? Change is merely the transition from one state to another—whether it be the movement of an object from one place to another or one temperature to another or even one speed to another. But it does not really tell us a great deal about the concept of speed, for example, to say that we began our travels in our new roadster, the Baguette, at a speed of 5 miles per hour and that our speed increased (or changed) during the course of our journey to 30 miles per hour. This statement merely tells us that at some point we decided to throw caution to the winds and stick our tongues out at the police sitting by the roadside and stomp the accelerator to the floor to feel the G-forces of 20, 25, and even 30 miles an hour. Such high speeds can be dangerous because it is almost impossible to take a deep breath when whizzing along the highway and passing by pedestrians and even elderly cyclists riding adult tricycles. This statement does not tell us, however, anything about the rate at which our speed was changing at any single point in time. In short, it says only that our speed did change—it provides us with no clues as to our rate of acceleration.

The rate of change is nothing to turn your nose up at because it can be of vital importance depending on the situation in which it occurs. Suppose that you are a passenger on a plane being piloted by the former stunt pilot "Happy Jack" Malloy. Happy Jack might wish to help you become more familiar with the rate of change concept by cutting the engines off and allowing the plane to free-fall. Or he might simply point the nose of the plane downward and open the throttle. The rate of change in this case would be the amount of feet per second that the plane falls while Happy Jack sings an old sailor's lullaby and all the passengers sob uncontrollably in hysterical terror. The plane might free-fall at 350 miles per hour or Happy Jack might choose to pilot it downward at 650 miles per hour. In either case, the passengers will see firsthand how different rates of change can affect their own sense of well-being. The free-fall might feel like a gut-wrenching plunge whereas Happy Jack's intentional effort to ram the plane at full speed into the top of a mountain would probably cause most of the travelers to scream in sheer heart-stopping terror. Of course any comparison of rates of change in this situation will likely be muddled by the deaths of all the passengers on board when the plane hits the ground but scientific experimentation cannot always proceed in a squeaky-clean laboratory.

The rate at which some variable—whether it be temperature, pressure, or volume—changes can have profound effects. Suppose you are one of those hardy souls who likes to jump into an unheated pool every morning and swim 50 laps to get your blood coursing through your veins before heading off to work. If you left the warmth of your feather bed and leaped into a pool which was cooled to just above freezing (33 degrees), you would detect a very rapid decrease in the temperature of the surrounding environment. It might be so profound that your heart would decide to stop beating, leaving you to flail away helplessly in the water before sinking to the bottom. You would probably avoid such dangers if your pool was kept heated at a more moderate temperature such as 70 degrees. Upon jumping into this pool, you would enjoy a very pleasant, warm sensation which would presumably not be due to an act of urination. The rate of change in the exterior temperature would be much less rapid and your prospects for a few more years of carefree joy on earth would be greatly enhanced.

Another example of the importance of a rate of change can be found in the courting ritual. Suppose that Reg likes Diane very much and desires to win her affections. When dealing with a demure woman such as Diane, the rate of change in the attention Reg showers upon her is crucially important. Reg may find that his prospects for success are improved if he follows the 10-step video tape on "Successful Wooing," which advises him to begin with casual dates such as lunches in public places with lots of witnesses and visits to weekend attractions such as mud wrestling tournaments and tractor pulls. Once he has shown Diane that he is a man of culture and learning, Reg can then intensify his romantic endeavors and take her out for an evening on the town at a fancy nightclub where scantily-clad dancers gyrate on table tops for patrons who stuff dollar bills into their panties. At this point, Diane should feel, according to the narrator on the video tape, that she has met the man of her dreams and succumb to his animalistic advances in a telephone booth outside the club. Obviously, the rate of change in Reg's wooing would be critical to Reg's winning of Diane's affections, occurring as it did in deliberate, measured increments. Had Reg chosen to follow his own instincts, by contrast, which involved a clever plan to show up at Diane's door at the start of their first date wearing nothing more than flippers, a mask, and a red arrow on his stomach pointing south, then the rate of change in the wooing process by Reg would have likely hit a critical overload and Diane would have found it oppressively intense. Indeed, it would have been too much too soon and she would likely have slammed the door in Reg's face or at least given him the address of her former girlfriend whom she knew to be a gun-toting psychopath.

Anyone who has ever held her breath and swum down to touch the bottom of a swimming pool will understand the importance of rate of change. If you go down too quickly without allowing the pressure on your ear drums to equalize, then you will feel a sharp pain. If you swim down to the bottom at a lower rate, however, then your ear drums should have adequate time to adjust to the increased pressure and your head will not blow up. This is a very important point to keep in mind when attending pool parties thrown by your boss and other people you wish to impress as it is considered a major *faux pas* to have one's head explode in another person's swimming pool. Here again, the rate of change plays the pivotal role in determining the outcome of the activity.

But we need to refine our idea about the rate of change even further to understand the difference between the average speed or average rate of change and the rate of change at a given instant. If we decide to run in the New York City Marathon and we finish that 26-mile race in 13 hours, then our average speed or average rate of change is 2 miles per hour. Such an unimpressive speed will probably not enable us to finish among the top ten runners or even in the same calendar day. But we may want to point out that our rate of change for the first 50 yards of the race was an impressive 20 miles per hour. Had we been able to maintain this pace for an additional 26 miles and 335 yards, then we would have finished the race nearly an hour under the existing world record. The fact that we did not have the "legs" to complete the remainder of the race should not detract from the fact that for a brief time we were ahead of the best marathon runners in the world and were running at a very impressive instantaneous rate of change (for 50 yards).

The difference in these two rate of change concepts is further illustrated by an experiment with a bullet and a gun. Suppose we convince our next-door neighbor (who is a bit of a dullard) to stand at the far end of the yard with a target in his hand so that we can illustrate the difference between instantaneous rates of change with a flourish. I could take aim at the target, pointing out that the bullet will be traveling at a very high velocity which, given the short distance involved, will approximate an instantaneous rate of change. I will also point out that I have only maimed a few people with a gun and that most of them have gone on to lead productive lives. Although my neighbor might be less enthusiastic about my plan than I, he would soon learn firsthand about instantaneous rates of change as the bullet ripped through the target. I would then point out that my tossing the bullet at the target with my hand would also have an instantaneous rate of change although it would clearly take the bullet a longer time to reach the target. There would certainly be less excitement associated with this type of demonstration due to the lack of danger and the extremely remote possibility of death or severe injury.

How quickly the bullet travels at a given instant is very important, particularly when viewed from my neighbor's perspective. Indeed, he may wish to engage in his own analysis of rate of change by seeing how quickly he can move to a safer location. But the important point is that the rate of change *at a given moment* can be of much more interest to us than the average rate of change or velocity of an object. The faster the rate of change in a given situation, the more profound the result. One need go no further than one's own home to drive this point home with the use of a bare foot and a furniture leg or door jamb. If you casually swing the edge of your foot against the furniture leg, thus hooking two or three toes around the leg, you may feel some discomfort but it will soon pass. But if you are more determined to see this rate of change analysis through and you swing your foot as hard as you can as though you were trying to kick a 50-yard field goal in a championship football game, then you will undoubtedly feel the dizzying rush of pure agony as you crush several of your toes. Aside from some brief concern as to how you will ever find sandals to fit your new "foot look," you might have enough clarity of thought as you hobbled down the stairs to consider the more profound implications of this rate of change concept. You would probably conclude that the more rapid rate of change did indeed create more profoundly painful consequences and that your experiment indicated that you should carry out your kicking drills at a more conventional location such as an athletic field.

The creators of the calculus were concerned with the rates of change at very tiny intervals of time. They did not want to deal with the average rate of change over a prolonged period of time such as a minute, an hour, or even a day. Instead they wished to shrink the time period being considered to an infinitesimally small amount so that they could analyze the rate of change at a given instant of time. You might ask why anyone would care about the rate of change at a given instant of time. But as we saw in our example of the bullet and the reluctant next-door neighbor, the rate of change at a given instant of time can be very important—particularly when one is talking about a bullet. I can certainly shoot a bullet through my neighbor without breaking into a sweat. But it is unlikely that I would ever be able to throw a bullet through my neighbor, even if I get a good running start.

But what do we mean by infinitesimal time? A scientist might explain it in technical terms as being an itsy, bitsy, teeny, weeny period of time. For those of us who are not so quantitatively adept, however, it will suffice to say that we are referring to fractions of a second or even the time which passes in the blink of an eye. Obviously, there is no single answer to this question because we can slice a second up into tenths, hundredths, thousandths, ten-thousandths or even smaller fractions of a second. This is not a very fulfilling career choice but it underscores the difficulty of trying to define a word such as "instantaneous." So whoever among the crowd is the most Napoleonic in temperament (and perhaps stature) may choose to decree that an instant is one one-hundredth or one one-thousandth of a second of time. For all practical purposes, anything smaller than a hundredth of a second will be imperceptible to our senses.

As most mathematicians will not be content with the metaphoric rants of mediocre poets regarding the fleeting passage of time, they will try to describe in mathematical terms what is meant by an instant. A mathematician may designate an instantaneous event by some symbolic description such as $t = 0$ where t is the reference to the time at which the interval occurs and the 0 is the number of seconds which have passed since the occurrence of the event. As no seconds have passed since the occurrence of the event, then we can conclude that the event is occurring at this very instant. Whether you are present when the event occurs can be very important. Suppose that your lover is experiencing the most intense, the most fervent physical sensations she has ever felt at $t = 0$ seconds. You would be very pleased if you were actually with your lover at $t = 0$ seconds; you would be very disturbed if you were across town at the instant $t = 0$ seconds and this very momentous occasion were occurring with another person. Even though you might learn of the event at $t = 3$ seconds when your lover called you up to inform you that she was leaving with her new friend to join a carnival in Barbados, the fact that you hopped in your car and drove across town like a crazed maniac to put an end to this diabolical plan would do you little good because the passage of time would only go forward (such as $t = 4$ seconds, $t = 5$ seconds, and so on). So even though you might arrive at your lover's apartment at $t = 340$ seconds to find a "good-bye" letter, the instant would have long since passed. Other than trying to race to the nearest train station or airport, you would have to admit that you would have to find a new lover with more conventional tastes in companions.

If we do not assign an arbitrary designation for when an event is occurring and thus define the point in time at which an "instant" occurs, then we find ourselves fumbling around for words, feeling much the way we would had we stumbled upon our adventurous lover and her "friend" at $t = 0$ seconds. How can we describe an "instant" in a purely descriptive, nonquantitative manner? We could stand on a street corner and try to use our command of the English language to say "I am touching my foot right now" and thus provide an insight into the concept of an instantaneous event. But does it really help us very much to watch someone touch his foot and simultaneously proclaim this obvious fact to the world? Well, if we are feeling depressed, then the sight of someone who is so pathetic that he must announce he is touching his foot to attract attention should bolster our own self-confidence. We would at least be able to point confidently to the foot-touching individual and feel ourselves to be a rung or two higher on the evolutionary food chain. But as far as our perceiving the act of touching the foot and the proclamation that the foot is being touched is concerned, we might find ourselves unable to define in any clearer terms what is meant by an instantaneous event.

Being clever individuals, we would probably decide to take a different approach in our investigation of the concept of instantaneous time such as using our sense of touch. I could turn the burner of the electric range in my kitchen up and wait for the coil to heat to a red-hot temperature, I could then demonstrate the passage of an instant of time by holding my hand just over the coil and pointing out that my hand is not actually touching the coil. I could then deftly place my hand on the coil and thus enjoy the scent of hickory-roasted flesh as well as the wholly unpleasant sensation of my palm incinerating. I would be able to determine that I had passed from the point of not having my hand on the red-hot coil to now having my hand on that very same coil. I would also have the additional tactile sensation of the heat from the coil passing through my hand at a speed that would seem to be instantaneous. Although I might believe that my having felt the pain of my decision to place my hand on the coil was instantaneous, there are some physiologists who would dispute the idea of an instantaneous sensation of pain. They would argue that the burning sensation would travel via the nerve endings in my fingers at a finite speed until it reached the appropriate part of my brain where I would be able to conclude in a rational manner that I had burned up the palm of my hand. So while I might feel that the shooting pain from my fingers had occurred as soon as I had placed my fingers down on the coil, these busybody physiologists would probably insist that my perception was incorrect and that a certain period of time, albeit a fraction of a second, would have elapsed from the time of contact to the time of my window-rattling screams of pain. So even though I might be willing to take foolish risks of personal injury in order to demonstrate the duration of an instant, I would probably find it a frustrating and, ultimately, unsuccessful endeavor.

This outcome would hold true even if I decided to flip a bedroom light switch off and on. The light itself would go off and on as I flipped the switch frantically and I might even attract the attention of a flock of rare whooping cranes which happened to be flying by my house at the time and cause them to ram themselves against the outside of the window. Even though the light might seem to go off and on in an instant, we would have to admit that the time for turning the light off and on was not truly instantaneous but finite. Of Course I might want to darken the lights and not continue this investigation any further once the members of the local animal rights group knocked on my door to try to find out why there are a pile of whooping cranes lying dead at my window. They might provide their own demonstration of the concept of instantaneous time by breaking down the door to my home and dragging me out by my feet so that I could be taken down to the local police station and arrested for cruelty to animals.

Because of the limitations on our sensory powers to perceive very small units of time which would approximate what we would consider to be instants of time, we may have no choice but to retreat to the mathematical symbolism we offered before and idealize this concept. Suppose that I have been training for a swimming meet for two weeks, having previously decided to resurrect my long-dormant athletic career. The seriousness with which I view this meet is underscored by the fact that I have cut my daily cigarette consumption down to four packs and am no longer pouring vodka on my pancakes. I have also greatly restricted my evening social life, making sure to be home in bed by 2:30 in the morning at the latest. With all of these positive changes bolstering my physical well-being, I am finding that I can make it down to the mailbox at the end of the driveway without slumping to the ground out of breath. In the pool, I am also experiencing newfound health and vigor—I no longer have to push off the bottom to make it to the other side. As you can imagine, all of this physical activity has made me feel like a new man. I have progressed to the point where I can swim more than one lap before needing oxygen and blood transfusions. This increased range is helpful as my training distances have advanced to the point where I can now swim as far as 100 meters—which, coincidentally, happens to be the distance I shall be racing. So on the day of the race, I will stand with seven other magnificent men on the starting blocks, hunched over in anticipation of the starter's gun, our faces unpleasantly close to our groins.

At the start of the race I dive into the water, arms whirling like a windmill, legs churning like a powerboat motor, swimsuit floating near the point of entry due to my having forgotten to tie the drawstring. But having never been one to let public humiliation get the best of me (and resisting the temptation to do the backstroke), I churn through the water with a vengeance, reaching the other side of the pool (the first 50 meters) in 30 seconds. Although I do a flip turn and catch my foot in the coping of the pool for a moment, I wiggle free and am once again racing back across the pool to the finish line. As legs and arms flail away in unison, I resist the temptation to stop to assist the competitor in the next lane who has apparently suffered a heart attack and sunk to the bottom of the pool. After all, one needs to look out for number one, particularly when one needs to get a towel around one's waist as quickly as possible. When I touch the other side of the pool, I wipe away the blood streaming out of my ears and wave to the adoring fans. My coach/accountant/therapist comes over and tells me that I completed the second lap in 36 seconds. As I have a hankering to toss a problem relating to the concept of instantaneous speeds around in my mind as I retrieve my swimming suit, I can consider the idea of average speeds and instantaneous speeds. My average speed in swimming the two laps was 33 seconds per lap. Now there may have only been a few seconds during the race in which I was actually swimming through the water at a rate of 33 seconds per lap. At some points I might have been swimming as fast as 28 seconds per lap such as when I managed to swing my arm over the lane marker and catch my finger in the waistband of the swimmer in the next lane. At other times, however, my speed might not have been quite so impressive, slowing down to perhaps as little as 39 seconds per lap such as when I dived into the water and noticed that I would have even less resistance than usual due to the extraordinary streamlining effect wrought by my having lost my swimming suit.

To determine my instantaneous speed at any given moment of time, I would have to follow the example of the mathematician and cut my 100-meter race into smaller and smaller intervals so that I could determine my speed at any given meter or even any given centimeter. The problem that arises is one with which mathematicians are very familiar. As the interval of time goes to 0, so does the interval of distance covered within that time. Ultimately, we end up in the rather preposterous situation in which we travel 0 meters of distance in 0 seconds of time, thus giving us an instantaneous speed of 0/0, a quantity which is undefined and does not contribute a great deal to our understanding of this concept. Although there were times during my training routines in which I traveled 0 meters for very long periods of time due to my having passed out from lack of oxygen, I could not claim to be able to make sense of traveling 0 meters of distance in 0 seconds of time any more than my colleagues in the sciences.

Now you could ask whether all of this concern with instantaneous speeds is worth the attention given to it by generations of mathematicians and physicists. Unfortunately for those who believe that we should devote more time to listening to the sage advice of personal psychics and seek the images of God in airbrushed liquor advertisements, there are a virtually endless number of physical problems in which the instantaneous speed of an object must be considered. These are not merely idle thought experiments performed by tweed-jacketed pseudo-intellectuals but important problems which manifest in almost every area of our lives. When one is dealing with instantaneous speeds, one might have to consider the speed at which the earth travels around the sun or the rate at which an artillery shell travels through the air before dropping on the heads of enemy soldiers. Although there is not much we can do to alter the velocity of the earth around the sun even though we might try to slow it down by loading the earth up with more and more of the huddled masses, we can use our knowledge of instantaneous speeds and instantaneous acceleration to build artillery shells that will actually travel far enough to hit enemy positions as opposed to our own troops.

Those lucky few who design artillery shells are concerned with the acceleration of the shell from the time it is fired to the time it hits its target. They need to be able to calculate the rate of acceleration of the shell at any given time in order to determine the trajectory of its path through the air. Knowing the rates of acceleration and deceleration as the shell rises and then falls toward its target is of crucial importance because it makes it possible to aim the shell. Firing artillery shells is not something that one can do on a lark because the shell may fall short of the target and hit the commanding officer's staff vehicle (with the commanding officer aboard) or it may sail over its target and detonate a passing garbage truck. Only after one has acquired some knowledge of the path traveled by the shell when it is fired from the cannon can the appropriate adjustments be made to improve the accuracy of the targeting itself. Here is where the scientific practice of making wild guesses comes into play. If a shell rises too quickly into the atmosphere and falls short of its target, we know that we must lower the angle of the cannon relative to the ground so that there is less of a vertical and more of a horizontal impetus to the path of the shell. Similarly, if a shell sails over a target, we know that we must increase the vertical and reduce the horizontal impetus to the path of the shell. Finally, if the shell travels straight up and lands on us, we will be dead so we will not need to concern ourselves with such banal questions.

Because almost all dynamic phenomena in nature can be viewed in terms of their instantaneous speeds and accelerations, the ability to calculate such quantities is of great importance to physicists and engineers alike. Mind you, physicists and engineers enjoy attending wild orgies where they can cavort in bathtubs of cheap wine with dozens of writhing naked bodies as much as anyone else but they also have their serious sides and are deeply interested in trying to understand the dynamic forces in nature—whether it be the movements of waves on the surface of a pond or the propagation of an electromagnetic field in space or even the invariable tendency of buttered bread to land on the floor buttered side down.

Until the 17th century, mathematicians and scientists had neglected to focus on the dynamics of rates of change. Although many individuals would contribute to the development of the calculus, it was the Englishman Isaac Newton and the German Gottfried Wilhelm von Leibniz who would independently develop the mathematics that would make it possible to calculate instantaneous rates. Although the methods and the nomenclatures they used differed, they both had the same goal in mind and created their respective mathematical edifices at about the same time. What is almost comical in retrospect is that the utility of the beautiful mathematics they had created separately became overshadowed by the chest-thumping nationalistic claims of priority made by English and German scientists and mathematicians, who were apparently no more immune to the nationalistic fervor sweeping through their lands at the time than any one else. Indeed, it was the self-evident elegance and universal applicability of the calculus which made it such an intellectual jewel; the nation which could claim its creation would certainly enjoy a great boost to its prestige even though one could not really claim a meaningful strategic advantage. After all, the fact that Newton might have invented the calculus a few years before Leibniz would not have given England a tactical edge if it had chosen to become embroiled in a war with Germany over an issue of vital national concern such as the relative health benefits of kidney pudding and bratwurst. Very few sailors in the Royal Navy would have actually been able to identify the word "calculus," let alone solve problems using its techniques. To the extent that they used technologies (such as ship cannons which fired cannonballs at opposing ships) which could be described with mathematical equations, they were guided by trial and error and not mathematical techniques. In other words, a sailor aiming a cannon relied on his own perception and experience in deciding the extent to which the cannon needed to be raised or lowered before firing at its target. If the gunner fired the cannon and blew a hole in the deck of his own ship, then his intuition would tell him that the cannon needed to be raised and that he should probably not count on a promotion to head gunnery officer any time soon. The gunner's decision to raise the cannon would not come from his having carried out a detailed mathematical calculation of the instantaneous rate of speed or acceleration of the cannonball at various points in its path from the ship to the target but instead from an educated guess as to how best to aim the cannon given the apparent distance of the target. In those days there were no electronic calculators and ships which became involved in pitched naval battles would often pull alongside each other and fire their cannons broadside. One did not have a great deal of time for thoughtful mathematical analysis when cannonballs were crashing in all around and the entire ship was engulfed in fire and smoke. Instead one had to load, aim, and fire as quickly as possible.

How can we better understand this beast that we call the calculus? Where should a thoughtful examination of this discipline begin? Both Newton and Leibniz were aware of the work done by Galileo Galilei on the acceleration of falling objects. Having some time on his hands, Galileo had climbed up to the top of the famous Leaning Tower of Pisa and dropped objects of varying weights from the top to see whether there was any difference in the rate at which they plummeted to earth. His effort to extend his testing to live humans was stymied by a lack of volunteers, but Galileo was able to draw some useful conclusions about the rates at which objects are accelerated toward the earth. In fact, he developed an equation which relates the distance an object falls to the time it takes to fall: $d = 16t^2$ where d is distance and t is time.

How does this equation work? If we climbed up to the top of the Leaning Tower of Pisa and began dropping heavy objects on the heads of the tourists, we would soon see that they dropped at the same rates of speed toward the ground. If we had good eyes and accurate watches, we would see that this rate could be quantified using Galileo's equation. It tells us that if we drop an object such as an original Rembrandt portrait, we can determine the distance it falls in a given time and *vice versa*. If $t = 1$ second, for example, then the distance the painting will fall in the first second is 16 feet. If $t = 2$ seconds, the distance increases to 64 feet. If $t = 3$ seconds, the distance balloons to 144 feet. Of course these figures assume that the painting is not caught in an updraft and sent sailing over the city of Pisa itself. Galileo himself formulated this equation based upon the assumption that the objects were falling in a vacuum. He was not actually able to test this equation precisely because he could not convince the Pisa authorities to build a big plastic bubble around the Leaning Tower of Pisa and pump all the air out so that he could carry out his experiments in a vacuum. Of course the lack of oxygen would have killed Galileo long before he could climb up the stairs to the top of the tower to begin his experiment so it is just as well that this plan was never put into action.

What is particularly interesting about Galileo's formula is that it all but screamed out Newton's theory of universal gravitation. But Galileo—unlike Newton—did not think of the rates at which objects fall as resulting from the mutual attraction of the falling objects and the earth on each other. His failing was that he did not see the accelerations of these objects toward the earth as manifestations of a universal attractive force but instead focused on the study of the phenomena of the falling bodies themselves. Galileo, of course, was not as fortunate as Newton, the discoverer of the law of gravitation, to sit under an apple tree and be hit on the head with a piece of fruit.

But Galileo, along with Kepler, was among the first scientists to approach his investigations of nature with the idea of being able to describe its dynamic processes using mathematical formulas. As shown by the many discoveries he made with his telescope, Galileo was that rare blend of theoretician and pragmatist—seeking to meld observation and theory with mathematics. Like Kepler, however, he always had to take into account the possible reactions of the Church to his investigations. Yet Galileo, for the most part, proceeded to follow his own mind and paid little attention to the dangers of testing the prevailing orthodoxy.

Galileo did not see the need for a new mathematics to explain rates of change. It fell to his successor, Newton, who always freely admitted his indebtedness to Galileo's work, to develop the calculus, beginning with the very same concepts contained in the $d = 16t^2$ formula described previously. Because mathematics consisted of sterile, static disciplines such as Euclidean geometry and algebra, it was necessary to consider the concepts of a "function" and the "limit of a function" in order to understand the more fundamental concept of rate of change.

When we say that one thing is a function of another, we are essentially saying that one variable is so intimately related to another that the value of the first variable corresponds in some way to the value of the second variable. This concept is really quite simple and manifests in every part of our daily routine. Suppose that I am trying to attract the attention of a beautiful woman who is standing next to me on the commuter train. The most direct, straightforward approach would be to say something to her, something so suave, so sophisticated that she would look at me with yearning eyes and pouty lips. We could think of my statement as one variable and her reaction to my statement as another variable. A comment on the weather being very pleasant would probably elicit a noncommittal nod of the head. A statement that her perfume reminds me of the odor of a livestock slaughterhouse would likely elicit a piercing glare. A declaration that she cannot possibly be wearing any underwear would likely cause her to slap me in the face. So her reaction could be said to be a function of my statements. Of course this is not an easily quantifiable relationship because we are dealing with nonquantifia-

Figure 9. Galileo Galilei. Courtesy of AIP Emilio Segre Visual Archives.

ble functions but it does illustrate the idea that one variable changes in response to changes in another variable.

In the physical universe, the equation $d = 16t^2$ expresses a functional relationship: The distance an object has fallen is directly related to the time which has elapsed from the time the object was first set

Figure 10. Sir Isaac Newton. Courtesy of AIP Emilio Segre Visual Archives.

into motion. The greater the amount of time which has passed, the farther the object has fallen. Newton himself saw that the distance traveled was a function of the time elapsed. He further extrapolated these thoughts into a general theory of gravitation, reasoning that the mutual gravitational attraction between two objects (such as the

Figure 11. Gottfried Wilhem von Leibniz. Courtesy of AIP Emilio Segre Visual Archives.

earth and the sun) was directly related to (or a function of) the distance separating the two objects. If Newton had a really large catcher's mitt and could have shoved the earth much closer to the sun, the gravitational attraction between the two heavenly bodies would have increased accordingly. The closer the earth to the sun, the greater the amount of gravitational force exerted by the earth and the sun on each other.

To take this idea one step further, we can assign arbitrary values to one variable—the so-called *independent* variable—which will automatically determine the corresponding value of the other variable—the so-called *dependent* variable. If we were to go back to Galileo's formula relating the distances objects have fallen and their elapsed times ($d = 16t^2$), we could assign any value (which we did do above when we plugged in the values 1, 2, and 3 for the variable t) to the time t (the independent variable) and we would thus obtain values for the dependent variable, distance d. So the distance d is a function of the time t or, alternatively, the distance d is dependent on the value assigned to the time t. To return to my strap-hanger example, we could say that the independent variable is the statement I make to the woman standing next to me and the dependent variable is the severity of the slap of her hand across my face. Mathematicians are not very interested in trying to express such amorphous relationships because, as we pointed out before, they are not easily expressed in quantifiable terms. But they do illustrate this idea that a function exists when the value of one variable is dependent on the value of another variable.

Those persons who drive the nation's highways with a burning death wish in search of concrete walls and oncoming vehicles into which they can ram their cars can easily appreciate this idea of dependent relationships. If the speed of the vehicle is the independent variable, then the amount of damage caused to the car upon impact is the dependent variable. Therefore, the cost of repairing the vehicle is a function of the speed at which it strikes whatever object happens to get in the way. This is not a perfect functional relationship because some cars will inexplicably suffer less damage following their collisions with concrete embankments traveling at higher speeds than other cars running into those embankments at lower speeds. But we would expect there to be a discernible relationship which tells us that the greater the speed at impact, the greater the damage to the car. This direct relationship does not continue upward forever but will instead reach a plateau. In this case, there is an upper limit to the amount of money needed to repair the car, regardless of the speed at

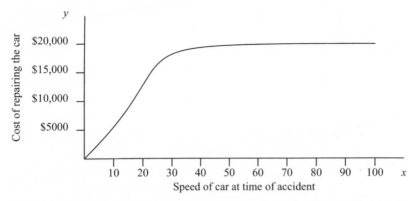

Figure 12. This graph illustrates that there is a direct relationship between the speed at which a car crashes and the costs associated with repairing the car. But at some point the costs reach a maximum which is equal to the replacement value of the car.

which it strikes the embankment. The reason for this is that the car will be destroyed at any number of speeds and the cost of repair (or replacement) will be the price of a new car. If we were to graph this relationship, we might express the dollars needed to repair on the vertical axis of a graph and the speed of the car at the time of impact on the horizontal axis, as shown in Figure 12. In the beginning there would be a rapid rise in the plotted points on the graph, but at some point we would reach an upper limit of repair costs at which time we would declare the car a total loss and specify its replacement cost. The line of the graph would then flatten out. No matter how hard we smashed the car into an embankment, the replacement cost of the car would remain the same and the graph would begin to approximate a line parallel to the horizontal axis. This would hold true even if we placed the car into an airplane and dropped it into a rock pit from an altitude of 7000 feet. Of course we might have to modify the vertical axis of our graph somewhat if the car that dropped from the airplane landed on a flock of rare dwarf sheep grazing at a petting zoo because the replacement value would now have to include the cost of catching a new flock of these sheep. But the important point is that these functional relationships do not always continue indefinitely but may gradually peter out over time.

As mathematicians are an impatient lot who like to get to the bottom of a matter as quickly as possible, they would not want to rely on an overly wordy description of a functional relationship. They would prefer to use a very concise mathematical equation to express a function. While those of us unskilled in mathematics might express the dependence inherent in any functional relationship by saying that "*y* is a function of *x*," those clever, tight-lipped mathematicians would find such a phrase far too wordy for their tastes. Indeed, they have their own special nomenclature for expressing the function dependence of one variable *y* upon a second variable *x*: $y = f(x)$. Note the clarity and conciseness of this expression—it is pouty yet full-bodied, profound yet simplistic. Indeed, it is one of the few mathematical equations where one can actually "read" the equation and understand it knowing very little about mathematics. The obvious importance of this equation is that it offers a mathematical description for any functional relationship and can be expressed graphically. One very simple example is the algebraic expression $2x = y$. We calculate the value of the function by assigning a value to *x* such as 2. If *x* is equal to 2, then *y* must be equal to 4. If we were to describe this relationship using our vaunted equation, we would write $y = f(2) = 4$. Being members of the visually-oriented generation, we would plot this point by going to our handy graph with the horizontal *x*-axis and the vertical *y*-axis. We would then move two marks to the right of the 0 of the *x*-axis to the 2 and then move upward four marks on the *y*-axis. This point, which we designate as (2,4), is one point on the curve which describes the function $y = 2x$. Well, we could break out the champagne and allow ourselves to become carried away by our brilliance in plotting this function but we would be sobered by the fact that we cannot really describe this function graphically yet because we have only one point. So if we want to continue our dalliance with this function, we will have to assign another value to *x* such as 3. Using our knowledge of fourth-grade arithmetic, we can determine that if *x* is equal to 3, then *y* must be equal to 6. This point will be designated as (3,6) and can also be marked on our graph by moving three marks to the right of the 0 on the *x*-axis and then upward six marks on the y-axis as shown in Figure 13. Now that we have two points describing this function, we also get a sense of the difference

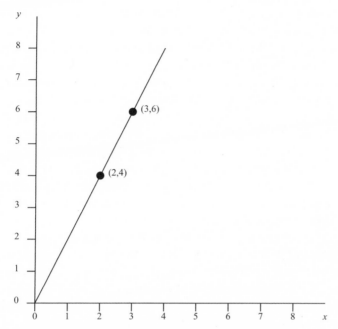

Figure 13. This graph describes the function $y = 2x$.

between being an independent variable (x) and a dependent variable (y). The independent variable x is selected arbitrarily and, once designated, automatically defines the dependent variable y. Thus the dependent variable has no say in the matter as to which position on the vertical axis it will occupy. The autocratic behavior of independent variables in general has not been of great concern to political scientists, many of whom care very little for the nitty-gritty details of the bruising world of differential calculus. But it is vitally important to mathematics because it is this bullying behavior of the independent variable which not only squelches the yearnings of the dependent variable to be free but also gives rise to the concept of the limit of a function. In other words, as the independent variable moves along the x-axis from the point where $x = 2$ to the point where $x = 3$, the dependent variable has no choice but to also move along the y-axis from the point where $y = 4$ to the point where $y = 6$. As the independent variable x approaches its limit or the value 2, the y variable approaches its limit or the value 4. Thus the limit of $f(x)$, as x approaches 2, is 4. This is the amount of "bend" in this function as the value for x inexorably determines the value of y as the x value approaches 2. This idea of the x variable moving toward the value 2 as the y variable moves toward the value 4 is what is meant by the limit of the function.

This concept of the limit is one which is tantalizingly simple yet frustratingly obscure. It can perhaps be better understood if we climb up to the top of a skyscraper and begin dropping lead weights. These weights are quite heavy and were quite a chore to lug up to the top of the building because the elevators were out of order. As a result, we were forced to drag the weights up 40 flights of stairs, risking heart attacks in the name of science. But once at the top of the building we could survey the world beneath us and begin to think about the best way to carry out this experiment so that if we hit any pedestrians on the head by accident, their deaths will not have been in vain but instead can be regarded as a vital contribution to the progress of science. We may also want to check the most recent list of countries which do not have extradition treaties with the United States so that any getaway which becomes necessary due to the unwillingness of the civil authorities to recognize the merits of our experiments will not be prematurely interrupted.

Once we have dealt with any potential legal problems which may arise from our experiments with dropping lead weights, we would also want to begin thinking about the ways in which such an experiment could illustrate the concept of the limit in a way that is helpful to us. Suppose that our building is 576 feet tall. We know that our lead weights will fall through the air at an ever-increasing speed but we know it is not an infinite speed (even though it may seem infinite to the unfortunate person who happens to cushion the lead weight's impact when it strikes the ground). As we have kept several world-renowned physicists on retainer so that they can answer any questions we might have even in the very late evening hours, we soon learn that our effort to understand better the concept of limits requires us to divide the path our falling weights will fall into a series of segments. We already know that the rate of acceleration of the lead weights will be governed by the equation $d = 16t^2$. This equation tells us that a lead weight dropped off the roof will drop 16 feet in the first second of its fall, 64 feet in the second second of its fall, 144 feet in the third second of its fall, 256 feet in the fourth second of its fall, 400 feet in the fifth second of its fall, and 576 feet by the sixth and final second of its fall. It is a pleasing coincidence that the lead weight will hit the ground exactly 6 seconds after we drop it off the top of the building. It is only with detailed planning and preparation that this type of precision is possible.

Now we could play it safe and content ourselves with calculating the average rate of speed of the lead weights over the course of their 576-foot fall to the earth. To arrive at this speed, we would merely divide the distance 576 feet by the time of 6 seconds to obtain an average value of 96 feet per second. This is an interesting fact which might impress a few grade-school students who happened to be passing through the classroom in which you were giving a lecture on their way to the cafeteria but it is not very useful information. Why? Well, our equation relating the distance an object will fall to the passage of time provides us with a very pointed clue as to the change which occurs in the speed of an object over time. The equation $d = 16t^2$ tells us that the speed of any object (if we neglect air resistance) will increase with time. The fact that the equation tells us that any given value of t will be squared in calculating the distance the object has fallen shows that the speed of the object must necessarily increase with time. If the speed is increasing with time, then there is very little point in viewing any calculation of the average speed of the object as being a logical stopping point in our investigation. It is a useful bit of information to have and may come in handy if you are ever subject, while attending a funeral, to a pop quiz on the calculation of average speeds of falling objects (as might be the case if the decedent had died in a skydiving accident after having mistaken the bookbag on his back for a parachute) but it does not address the fact that falling objects increase in velocity with time.

Calculating average speeds would be more helpful if all objects fell to the ground at constant rates of speed. But we live in a quirky universe in which the speed at which any object falls varies with time. As we have seen from our trusty equation $d = 16t^2$, the speed increases with time. The mathematical problem posed by this fact is that we cannot point to any single period of time during the fall of an object to earth at which the speed of the falling object remains unchanged. Indeed, if we were to return to our falling weight example above, we would see that each second saw an increase in the distance through which the weight fell. We can break each second down into half-seconds, quarter-seconds, tenths of seconds, hundredths of seconds, and so on, in the hope of finding some temporal segment in which the speed is unchanged. However, our search will be in vain, at least to the extent that our watches and clocks can tell because we will find ourselves splitting our time segments into ever-smaller pieces until we get to a point where we simply cannot measure the segments accurately.

As we are all very busy people with many important things to attend to during our daily routines, we do not have the leisure time to do endless calculations of the average speed of the lead weights at any given point in their journey downward toward the hapless pedestrians below. Indeed, we would have to ask ourselves why we would really have any interest in bothering to calculate the speed of a lead weight 2.457 seconds or 3.453 seconds or 4.121 seconds after being dropped from the top of the building. There does not appear to be much point in doing such a calculation because it involves making an arbitrary selection as to an exact moment in the journey of the lead weight downward and deriving a number. On the other hand, we might be very interested in calculating the instantaneous speed of the object when it hit the ground or the pedestrian or the decorative flower beds adorning the front entrance of the building. In making this calculation, we would idealize the experiment to some extent in that we would ignore any air resistance, passing birds who might venture into the path of the falling weight, alien aircraft which might choose to aim their tracking beams at the objects to test their weapons before beginning their invasion of the earth, and any other factors which might alter the calculation. In short, we would pretend that we were dropping the weights in a vacuum because that would enable us to ignore all of these extraneous factors which might affect the time it takes the falling object to reach the ground. Pretending can be a virtue in science as well as childhood because it simplifies almost any kind of experiment a great deal by allowing us to reduce the number of factors which must be taken into account in carrying out our investigations and making our calculations. It is also considerably less expensive than building a gigantic 600-foot-tall vacuum chamber around our 576-foot building and timing the falls of the lead weights to the ground.

So how would we go about determining the instantaneous speed of a lead weight once it was dropped off the building? We would want to focus on the final second of the descent of the weight because that will give us a fairly good idea as to the speed of the weight when it hits the ground. We know that the weight will have dropped 400 feet by the end of the fifth second and 576 feet by the end of the sixth second. The distance that the weight will fall in the last second is 176 feet (576 − 400 = 176). This tells us that the speed of the weight in the final second is 176 feet per second. This is a very impressive speed and the plummeting weight will certainly leave a lasting impression (several inches deep) upon the crown of any unfortunate person who happens to be in the way. It is also the first step in our effort to determine the instantaneous speed of the weight when it strikes the ground.

Going back to our trusty acceleration formula, we then find that the falling weight traveled 84 feet in the time interval 5.0 to 5.5 seconds and 92 feet in the time interval 5.6 to 6.0 seconds. Because we have supremely lucid intellects, we can see by the additional 8 feet covered in the second half of the sixth second of the lead weight's fall to earth that it is still accelerating at the same constant $d = 16t^2$. So we have split that last second into two parts and determined that the weight will be falling at an average rate of 184 feet per second in the last half second of its fall. If we split the last half second of the fall into two halves (the final two quarters of the last second), then we move even closer to approximating the speed of the weight at impact. Indeed, we find that the weight will fall 45 feet in the next-to-last quarter second of its fall and 47 feet in the last quarter second of its fall. The per second speed of the last quarter second is equal to 188 feet per second.

As we all have many forms of mindless entertainment calling us away from these mathematical calculations, we will now actually begin making a point so that some valuable lesson can be drawn from this discussion. The perceptive reader will recognize that we are steadily shrinking the time interval in our lead weight model and are thereby more closely approximating the instantaneous speed of the weight when it hits the ground. In the final tenth of a second, the weight will travel about 19.04 feet so that its per second speed at that point will be equal to 190.4 feet per second. In the final hundredth of a second, the lead weight will travel 1.9184 feet or at a speed of 191.84 feet per second. But because most of the people who read this book crave greater precision in their mathematical calculations, we will carry this process out several more times to fully satisfy our anal-retentive tendencies.

To continue with the saga of the falling weight, we then calculate that the distance traveled by the falling weight in the final thousandth of a second will be about 0.191984 feet which would be equal to a per second speed of 191.984 feet. For those who have been wondering with nervous anticipation about the speed of the falling weight in the final ten thousandth of a second, however, we can now say that the distance the falling weight will travel in the final ten thousandth of a second will be equal to 0.01919984 feet, which is equal to a per second rate of 191.9984 feet.

But some of you are already wondering about the spirit of a book that does not immediately try to calculate the speed of the falling weight in the final one-millionth of a second of its fall. Well, we are not the types to hide from a challenge to our mathematical prowess so we shall take the plunge and report that the distance traveled by the falling weight in the final one-millionth of a second of its fall from the 576-foot-tall building is approximately 0.000192 feet which would be equal to 192 feet per second. As our calculator is one of the old-fashioned models that only carries its calculations out to six decimal places, we will find that it will necessarily round off its figures to six decimal places. So even though the distance traveled in the last one-millionth of a second is infinitesimally smaller than 0.000192 feet and the speed is 192 feet per second, we will treat it as equal to that amount because we can still make our point about the limiting value of the speed of the falling weight. What is the limiting value? We simply mean that as the time intervals are chopped into tenths, hundredths, thousandths, ten thousandths, and even millionths of a second, we come ever closer to approximating the speed of the weight as it hits the ground, which, in this case, is 192 feet per second.

This series of calculations whereby we were able to determine the speed of a falling lead weight dropped from a 576-foot-tall building in the final one-millionth of a second of its journey down the side of the tower illustrates the vast predictive power of even the most innocuous-looking equations. Let us remember that we have used only a single equation—the durable $d = 16t^2$—and it has provided us with the means for calculating the acceleration of an object falling to earth from any height for any interval of time to any desired degree of precision. The only limits are those of our own patience and sanity as one can go mad splitting intervals of seconds into smaller and smaller bits of time. Of course some people like the fact that they can calculate the speed of a falling object in the final one-billionth or one-trillionth or even one-quadrillionth of a second. But we have seen by the concept of the limiting value that there is steadily less to be gained by insisting on ever greater precision because the improvements in precision become negligible as we split our seconds into very small increments. From the standpoint of our own example with the falling weight, we can assume that it is falling at the rate of 192 feet per second at the time it hits the ground even though the actual amount may be 191.999999999999 feet. It simply does not make sense for us to worry ourselves with such puny discrepancies, particularly when we can spend our leisure time participating in a virtual smorgasbord of social activities such as wine tastings, nude volleyball, and spirited burnings of public buildings. But this continual splitting of time into ever smaller intervals provides us with a conceptual snapshot of the calculus—the approximation of instantaneous speeds at instants of time.

Now that we have used our equation to split up the final second of the fall and to calculate the speeds traveled by the weight at various intervals within that final second, it remains for us to do what good mathematicians do and adorn our equation with imposing algebraic symbols. The distance traveled by the falling weight can be represented by the algebraic equation $y = 16x^2$, which is merely a dressed-up version of our original equation $d = 16t^2$. All we have done is to substitute the variables x and y for the original variables t (time) and d (distance), respectively. But we need to take this nifty nomenclature one step further and describe the changes in distance fallen and time elapsed in a shorthand way. To represent the ratio of change in distance over the change in time, we borrow the Greek letter "delta" which is represented by a small triangle (Δ) and place Δy, the change in distance, over Δx, the change in time.

Now we are well on our way to creating some of those very ominous equations which appear in all calculus books. We express the functional relationship between the distance traveled and the time elapsed with the expression $\Delta y = 16 \, \Delta x^2$. Over the course of the 6 seconds which elapse during the fall of the lead weight from the top of the building, we can express the functional relationship as $576 = 16(6)^2$. But for any time less than 6 seconds during the fall of the weight, let us now think of Δx and Δy as those last little time and distance intervals just prior to impact and express the incomplete fall of the weight as follows: $(576 - \Delta y) = 16(6 - \Delta x)^2$. We can then simplify this equation as follows: $576 - 16(6 - \Delta x)^2 = \Delta y$. Although we could skip ahead several steps in this simplification process, we will instead lower our shoulder and plow ahead and spell out each step in the simplification sequence:

$$576 - 16(6 - \Delta x)^2 = \Delta y$$
$$576 - 16(36 - 12\Delta x + \Delta x^2) = \Delta y$$
$$576 - 576 + 192\Delta x - 160\Delta x^2 = \Delta y$$
$$192\Delta x - 16\Delta x^2 = \Delta y$$

The equation $192\Delta x - 16\Delta x^2 = \Delta y$ expresses the last little distance Δy prior to impact in terms of the last little time Δx prior to impact. To find the average speed, we divide $\Delta y / \Delta x$, which is equivalent to $192\Delta x - 16\Delta x^2$ divided by Δx. Thus, $\Delta y / \Delta x = 192 - 16\Delta x$.

You are no doubt extremely impressed with how we have managed to gunk up what formerly appeared to be a fairly concise and straightforward explanation of the functional relation between distance and time in this narrative. But this equation tells us that as we move closer and closer to approximating the speed at which the lead weight was falling when it hit the ground (the speed at which the time elapsed is equal to 6 seconds) the limit of the ratio $\Delta y / \Delta x = 192 - 16\Delta x$ which is 192. If we were to translate this statement into everyday English, we would say that as the Δx moves closer and closer to zero, the value of $\Delta y / \Delta x$ (or $192 - 16\Delta x$) similarly moves closer and closer to 192. So this equation tells us that 6 seconds after the lead weight is dropped off the building, its instantaneous speed will be equal to 192 feet per second. The rate of speed of 192 feet per second is the *limiting* value of this function. So we can borrow the verbiage of the calculus professor and say that the ratio $\Delta y / \Delta x$ approaches a limit and that its limiting value may be expressed as dy/dx. To express it in more concise terms, we say that as the value of Δx approaches 0 (gets closer to hitting the ground), the limit of $\Delta y / \Delta x$ is dy/dx.

Now we are getting somewhere! Unfortunately, we do not know whether the light at the end of the tunnel is an oncoming train. But we should take a deep breath and not be too intimidated because we have come so very far from our starting point. Indeed, we have managed to get a hand (or at least a finger or two) on the underlying concept of differential calculus. Yet our journey is not complete and we may still have to dodge a few falling rocks and mud slides before we get to a warm and cozy room for the night. But we at least have a general idea as to where we are headed and it now remains for us to bring in the mathematical symbolism that so terrifies most of the general population. Even though it appears to be forbidding, it is no different than another language which uses a different type of alphabet. Each of the symbols has a meaning which can be understood with a minimal amount of effort. We should remind ourselves that these symbols represent ideas which we have already discussed and wrestled with in trying to understand better the concepts of function and rates of change. The important point to bear in mind is that change is expressed in terms of the ways in which the dependent variable y changes in response to changes occurring in the independent variable x. The limiting ratio of the changing variables $\Delta y / \Delta x$ is expressed as Δx approaches 0 and for every value arbitrarily assigned to Δx there is a corresponding value for Δy.

When we say we are calculating the limit of a function such as our determination that the instantaneous speed of the lead weight would approach 192 feet per second as it struck the ground after being tossed off a 576-foot tower, we are saying that we are determining the derivative of that function. The derivative has nothing to do with movie sequels or knock-off clothing products but is merely another expression for the limit of a function. We are concerned with the rate of change of the function which, in the case of the falling lead weight, was the rate at which the speed of the object increased as it plummeted toward earth. When we speak of the derivative, we are speaking of the limit of the rate of change of that function, which, in our case, was the speed of the object as the time to impact approached 0.

So what if we want to determine the rate of change of a function $y = f(x)$ at a given point x_1? Well, we will merely return to the techniques we developed earlier in this chapter by taking smaller and smaller intervals to obtain more accurate approximations to the instantaneous rate of change at the point x_1. In other words, we will let Δx approach 0 and all will be well with the world. We dust off the equation $[f(x_1 + \Delta x) - f(x_1)]/\Delta x$ which allows us to approximate the instantaneous rate of change at x_1. This equation thus tells us that as Δx moves closer and closer to 0, this equation approaches a limiting value which we represent with the expression dy/dx and call the derivative of the function $f(x)$ at the point x_1. But we are global thinkers and we should point out that we did not go through all the trouble to construct an equation that works only for a single point in the function $y = f(x)$. You will be very pleased to know that you have gotten top value for your book-buying dollar and that this equation $[f(x_1 + \Delta x) - f(x_1)]/\Delta x$ may be used to calculate the derivative of the function $f(x)$ at any point—not just our arbitrary point x_1. So the derivative may represent the instantaneous rate of change of the function as the independent variable x takes on any quantitative value. As a result, it can be used for any arbitrarily chosen point such as x_1, x_2, x_3, and so on.

You might be saying this is all quite fascinating but still amounts to a lot of gibberish. This is admittedly a somewhat abstract concept which we can try to illustrate if we return to the field of geometry and imagine a two-dimensional graph with an x-axis and a y-axis. With the flourish of a young Picasso, let us draw a curve which drops like the side of a crescent moon so that it resembles the edge of a circle as

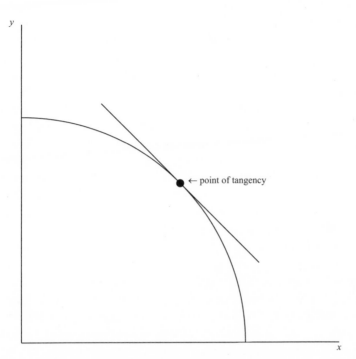

Figure 14. A graph illustrating the point of tangency—that point at which a line touches a curve at a single point without cutting through it.

shown in Figure 14. This curve shall be our line that we shall analyze. Now we know from our schoolyard experiences that a basketball (when properly inflated) comes into contact with the ground on only a very small area of its surface. If a basketball is underinflated, of course, it does not bounce at all but merely flops on the ground, much like older big league baseball players trying to steal home. But if we set the basketball on the ground, the ground touches it (or is tangent to it) at only a very localized area. Now we can covert this three-dimensional image to a physical counterpart of our two-dimensional graph with its artistic curve by imagining ourselves taking the basketball and the pavement upon which it is resting and slicing through them as though we were cutting a vegetable. If we continue to slice narrower and narrower layers, we will ultimately end up with a real-world counterpart to our idealized graph. Our examination of the slice of basketball will show that it touches the slice of pavement only along a very limited portion of its circumference. Because the edge of the pavement is perfectly flat (assuming it was initially built by honest paving contractors who do not periodically drop their former executives into the East River), it is said to be tangent to the circumference of the ball at the point where it touches the ball. To return to our curve (which we almost sold to an abstract art museum), we recall that we have a curve that resembles the edge of a circle, starting fairly high up on the y-axis and then sweeping downward in an arc out along the x-axis. We then draw a straight line so that it touches the arc at a single point without cutting through it. This point of tangency defines the slope of our curve at that given point.

What do we mean by the slope? Surely, it has nothing to do with snow skiing or mountain climbing. No, not really. But the slope of a line is merely the change in its vertical height over a given length of horizontal distance. Before delving into the mathematics of the slope of a line, we can better understand the notion of slope by pretending that we are standing on a gently rolling hill. If we run down the hill, we could say that our path was negatively sloped because our vertical height diminished as we moved along the horizontal axis. But if we ran back up the hill, we could say that our path was positively sloped because we had increased our vertical height while moving along our horizontal axis. If we test this analysis several more times, running up and down the hill at full tilt, we will begin to experience symptoms such as dizziness, shortness of breath, and numbness in the arms, all reflective of the more profound experience of cardiac arrest. Given the potential problems that such an experience can pose to one who is trying to carry out careful mathematical analyses, it is perhaps better that we confine ourselves to the theoretical world of lines and curves.

So let us return to the curve on our graph. Let us further imagine that there are two separate points on the curve, which we shall identify as P_1 and P_2 as described in Figure 15. There is nothing extraordinary about either point; their only distinction is that they lie at different positions along the curve. The differences in their x-axis

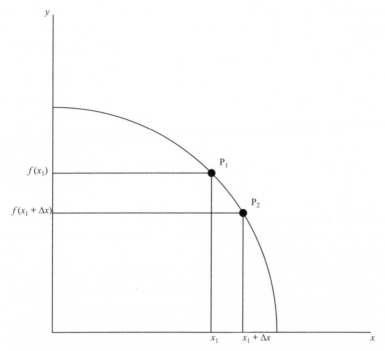

Figure 15. An algebraic illustration of changes in position along a curve.

positions may be described by x_1 and $x_1 + \Delta x$, where Δx is the spatial difference between P_1 and P_2 along the horizontal axis. The differences in the y-axis positions of P_1 and P_2 may be described using the equation of the curve $y = f(x)$ and are $f(x_1)$ and $f(x_1 + \Delta x)$. The change in the slope (the quality of "uphillness" or "downhillness") is obtained by dividing $f(x_1 + \Delta x) - f(x_1)$ by Δx. As the quantity Δx moves closer and closer to 0, the point P_2 moves along the curve toward point P_1. After all, the only thing separating the two points P_1 and P_2 along the curve is Δx. So the more we reduce Δx, the more we reduce the linear distance separating P_1 and P_2 along the curve.

What is this obsession with lines and curves? Although we would like to think that this book is multidisciplinary, this discussion of the curve representing the function $y = f(x)$ has nothing to do with latent sexual obsessions or subliminal advertising techniques but instead is merely our effort to draw pictures of a certain type of mathematical concept. But the slope of the curve at any given point provides us with a geometrical version of the derivative in that the slope indicates the rate at which the line is rising or falling whereas the derivative provides a quantitative rate of change. *Thus the slope of the tangent at any given point along a curve is equal to its derivative because the slope amounts to a geometrical rate of change.*

Now it is a good idea to step back and consider where we are going in our journey through the calculus. Newton had invented the calculus to explain the rates of change of dynamical entities such as the motions of particles with respect to time and, as a result, was primarily concerned with its utility in physics. Leibniz, by contrast, was more concerned with the abstract mathematics of the calculus and its general applications to knowledge as a whole. Neither man, however, invented calculus with the intention of manipulating the vulnerable and innocent minds of young students. They did share a common desire to develop a mathematics that would express the rates of change of various things that affect our lives such as the rate at which our weights careen from skeletal to obese, the rate at which we age, and the rate at which our bank balances rise and fall. If we are fortunate enough to exercise our right to travel as free-thinking citizens and load our automobiles up with children and luggage and spouses and cross the continent while pretending to be oblivious to the constant fighting and screaming going on in the back seat, we will have ample leisure time to consider the rate of change of our distance from our destination (our speed) at any given moment. As the screeching and the crying approach decibel levels that would drown out a jet plane engine, we may also consider the feasibility of adjusting the rate of change of the speed (the acceleration) so that we might get to our destination as quickly as possible and find relief in a separate hotel room with an intravenous morphine drip bag. This terrifying example, which nearly every parent who has ever set off on a month-long automobile vacation can appreciate, requires that we be aware of two different rates of change—speed and acceleration.

This example is not meant to trivialize the importance of being able to analyze rates of changes because such examples do manifest in almost every part of our daily lives. Whether we are dealing with the average length of time over which a tank of gasoline will power our car or the rates at which various cells divide or the rates at which radioactive nuclei decay or the rate of the flow of the money supply or even the rate at which we receive money as employees or proprietors, we must deal with rates of change. But we also need to see some of the ways in which derivatives relate to functions. How might we deal with a derivative that has a slope of zero?

First of all, we would need to understand what we are saying when we refer to a slope of zero. Remember that the slope of a line is a geometrical representation of its rate of change. If we set $f(y) = 2$, we are in essence saying that the values we plot along the vertical or y-axis (no matter where we are on the horizontal axis) are 2. This means that our graph will be a flat line which passes through the 2 on the vertical axis and extends indefinitely far in both directions, two units above and parallel to the horizontal axis. When there is no slope in the line (as is the case here), the derivative is zero because the rate of change is zero. How might we put some flesh on this idea? We could imagine Lance Redletter, world champion marathoner, hoofing it through the streets of Boston in his quest for another championship. But because we needed to have Lance assist us with our demonstration of derivatives of constant functions (unchanging slopes), we welded his shoes together, forcing him to hop along the race course. Because Lance can only hop at a steady 3 miles per hour, he does not reach a very impressive speed or, for that matter, stay with the race front-runners. But we would show Lance's speed as a constant 3 miles per hour by drawing a straight line through the vertical axis of a graph in which we would plot his speed against a horizontal axis reflecting time elapsed or distance run. In either case, the line graphed would be flat, underscoring Lance's inability to exceed the formidable 3-mile-per-hour barrier. To borrow the language of the mathematician, we would say that the derivative of a constant function is equal to zero. Of course this example is rather unrealistic because Lance will eventually get tired of hopping up the hilly streets of Boston and his speed will gradually slow down, no matter how adept a hopper he may be. Indeed, we would expect that the line would begin to slope downward, first almost imperceptibly and then more profoundly, as fatigue and, later, exhaustion, set in and Lance's speed declined from 3 miles per hour to 2 miles per hour and then to 1 mile per hour. Or we might find that the line came to an abrupt end at some point, thus signaling to us that Lance had keeled over in the street, having been struck down by a tourist bus driven by a near-sighted driver too proud to wear corrective eyewear.

But the use of derivatives is very common and mathematicians have developed shortcuts to enable those of us with full lives to cut through the chaff and calculate derivatives with little more than a bit of multiplication. Suppose that we have untied Lance and given him a transfusion of oxygenated blood laced with amphetamines so that he is far more vigorous than any runner who has ever raced along the Charles River. We then begin to time his progress and make a note of the distance traveled with the passage of each minute of time. When the time is equal to 0 seconds, Lance's position is also equal to zero because he has not yet begun to run. After 1 second has passed, Lance has run 6 feet. After 2 seconds, he has run 24 feet, after 3 seconds, 54 feet. We would represent this function as $y = 6x^2$, where x would be equal to the number of seconds being considered in our analysis and y the number of feet Lance has traveled since beginning his journey. If we graphed this line, we would find it to be very steep, beginning at the point at which the horizontal and vertical axes meet (both the values for x and y are 0). The next set of values would be $x = 1, y = 6$, the third set of values $x = 2, y = 24$, the fourth set of values $x = 3, y = 54$, and so on. Our mathematician friends would say that the function $f(x) = 6x^2$. This function shows in a very neat and compact way the way in which the time elapsed relates to the distance traveled by Lance from the beginning of his run. Assuming that Lance continues to run at a rate which corresponds to the pace described in the function $f(x) = 6x^2$, then he will have traveled 150 feet in 5 seconds, 216 feet in 6 seconds, and 294 feet in 7 seconds. At a time of 7 seconds, Lance, still presumably flush with his oxygen-rich, drug-laden blood, will have run the length of a football field. This would be extremely impressive because world-champion sprinters are seldom able to run a 40-yard dash in less than 4 seconds. Even if they could maintain such a torrid rate of speed, they would still take at least 10 seconds to run the length of an entire football field. Our friend, Lance, by contrast, will have covered that distance in only 7 seconds, which means that he would be running about 30 percent faster than the fastest human beings in the world. But Lance would admittedly be the beneficiary of our own special racing techniques. Even though Lance might find his heartbeat racing at three times its normal pace, he would continue racing across the land, his distance from the starting point continuing to increase with each second in correspondence with the function $f(x) = 6x^2$.

Because we are serious about the work we do, we would not want to tell Lance to ease his relentless pace, even though he might be sweating profusely with his tongue hanging down to his elbows and his eyes glazed and blood streaming out of his nose and ears. After all, we scientists cannot allow our analysis of our data to become colored by concerns that Lance might drop dead at any moment. So we would simply mix ourselves another drink in the back compartment of the limousine we are using as a chase vehicle and adjust the air-conditioning so that we do not feel too chilly and instruct the driver to honk the horn every time we see Lance begin to stumble or start to slow down.

At this point, we can say that we understand the relationship that is being expressed by $f(x) = 6x^2$. Each tick of the clock is accompanied by Lance moving farther away from the starting point. Each tick of the clock is also accompanied by Lance complaining about some imaginary ailment such as dizziness or blurred vision or aching joints or irregular heartbeats. But we can all be very pleased that Lance is continuing to maintain the torrid pace required by this cold, heartless mathematical function which turns a deaf ear toward Lance's suffering.

In summary, the function gives us the relationship between time elapsed and distance traveled and thus enables us to calculate the speed at which our plucky runner is racing. But what is the role of the derivative in this inspiring story? Quite simply, the derivative represents the rate of change in Lance's speed from second to second. It expresses the additional increase in feet covered by Lance with the passing of each second.

Although we could trot out a rather imposing array of equations showing how we obtain the derivative of the function $f(x) = 6x^2$, we could save ourselves a lot of time by using a short-cut method which merely requires us to multiply the exponent 2 by the coefficient 6. So the derivative of the function $f(x) = 6x^2$ is equal to $12x$. We express the derivative as $f'(x) = 12x$. Clever readers will notice the addition of the ' after the f, which provides mathematicians with a concise way of distinguishing derivatives from their original functions. The derivative $f'(x) = 12x$ means that Lance's distance from the starting point will increase by 12 feet with each tick of the clock. The rate of increase is 12 feet per second. This can be shown quite easily if we return to the values we had obtained for the dependent variable y in the function $f(x) = 6x^2$: At time $x = 1$ second, $y = 6$ feet; at time $x = 2$ seconds, $y = 24$ feet; at time $x = 3$ seconds, $y = 54$ feet; at time $x = 4$ seconds, $y = 96$ feet; at time $x = 5$ seconds, $y = 150$ feet; and so on. One could carry these calculations out for 10, 20, or even 50 pages but we do need to make a point here. First, we need to point out that once we get this function under way, the difference between the distance traveled between the first and second second is 18 feet (24 feet − 6 feet), between the second and third second 30 feet (54 feet − 24 feet), between the third and fourth second 42 feet (96 feet − 54 feet), between the fourth and fifth second 54 feet (150 feet − 96 feet), and so on. The one constant as each second passes is the rate of increase in the amount of distance covered, which is equal to 12 feet. We see that 30 feet is 12 feet larger than 18 feet, and that 42 feet is 12 feet larger than 30 feet, and so on. As a result, we can say with great confidence that Lance's rate of acceleration as he continues to huff and puff his way toward immortality is 12 feet per second. Fortunately, we do not need to go through all this trouble to determine the acceleration of the function $f(x) = 6x^2$ because we can obtain this figure—which we call the derivative—by following the step noted previously and deriving $f'(x) = 12x$. So the derivative gives us the rate of increase of this function which, in this case, is Lance's rate of acceleration.

What if we have a more complicated function such as $f(x) = x^2 + 2x + 2$? This looks very scary! If this function expressed the distance traveled by Lance over time, we might wonder at first glance whether Lance was doing somersaults along the way. or possibly running on his hands. After all, this function has two additional terms. But we need not allow its apparent complexity to scare us because we are brave and hardy souls who dare to venture into the darkest recesses of mathematics. If we follow the same process we used before, we will soon see that it is not as intimidating as we may have first believed.

Suppose that we have managed to convince Lance to take a shot at running at a speed which can be described by the function $f(x) = x^2 + 2x + 2$. This would not be a small accomplishment because of Lance having nearly succumbed to heat exhaustion in his prior race. But we are fortunate in being able to call upon our quack company doctor who can give Lance another dose of "groovy blood" and send him on his way. Even before Lance begins running, however, where time $x = 0$ seconds, we see that Lance is already 2 feet ahead of the game because $f(x) = 2$ where $x = 0$ seconds. Since the time has not begun to elapse, Lance could not have begun to run. The fact that he is 2 feet from the starting line merely means that he started running his race 2 feet in front of the original starting line. So throughout the entire race, Lance's position along the course will be shifted forward an additional 2 feet.

What are the values for the function $f(x) = x^2 + 2x + 2$? At $x = 0$ seconds, Lance is 2 feet from the starting line; at $x = 1$ second, Lance is 5 feet from the starting line; at $x = 2$ seconds, Lance is 10 feet away; at $x = 3$ seconds, Lance has raced 17 feet; at $x = 4$ seconds, Lance has traveled 26 feet; at $x = 5$ seconds, Lance is 37 feet from the starting line, and so on. We know that he is accelerating (at least while the "groovy blood" is able to offset the symptoms of his heart attack) because we see that he is covering a steadily greater incremental distance with each passing second. This is clear from the above numbers which show that in the first second, Lance covered 3 feet more than he had covered in the previous second. In the second second, he covered 5 feet more than he had covered in the first second. In the third second, he covered 7 feet more than he had covered at the end of the second second. In the fourth second, he covered 9 feet more than he had covered at the end of the third second, and so on. We might say to ourselves that we are witnessing something truly profound or we might simply be hearing the rumbling of our stomachs calling for a meal. We see that each distance covered with each passing second increases by 2 feet. What does this tell us about Lance's rate of acceleration with each passing second?

Acceleration? As we recall, acceleration is obtained by finding the derivative of the function. We have not yet forgotten how to find the derivative. Using our handy dandy method for finding derivatives (rates of acceleration), we see that the derivative of the function $f(x) = x^2 + 2x + 2$ is equal to $f'(x) = 2x + 2$. Hmm. This function and this derivative look much more forbidding than the first function $f(x)$ $= 6x^2$ and its derivative $f'(x) = 12x$. Never fear! The best way to approach this problem is to remember that even the greatest of mathematicians put their pants on one leg at a time and that we merely need to plug in different numbers for the x variable to determine the amount of distance traveled over time. Once we have arrived at a table of values, we can then determine whether the derivative we have calculated correctly expresses the rate of Lance's acceleration.

For the function $f(x) = x^2 + 2x + 2$, the derivative is obtained by multiplying the exponent "2" by x and dropping the x off the middle term, giving us the derivative $f'(x) = 2x + 2$. So we are saying that Lance's rate of change of position, or acceleration, can be expressed as $f'(x) = 2x + 2$ which means that as the value for x changes (where x = 1 second, 2 seconds, etc.), the rate of acceleration (or the slope of the line if we feel the need to give in to our artistic urges and draw a graph) will change accordingly.

Now we can go ahead and beat this point into the ground. The distance traveled by Lance between his starting point (which is 2 feet in front of the starting line) and the first second is 3 feet (5 feet − 2 feet). Not surprisingly, the distance Lance huffs and puffs between the first and second second is 5 feet (10 feet − 5 feet); the distance Lance totters between the second and third second is 7 feet (17 feet − 10 feet); the distance Lance stumbles between the third and fourth seconds is 9 feet (26 feet − 17 feet); and, finally, the distance Lance crawls between the fourth and fifth seconds is 11 feet (37 feet − 26 feet). The derivative $f'(x) = 2x + 2$ tells us that the incremental change in the distance covered by our brave runner with each passing second will change by 2 feet.

There is a rule for finding derivatives at work here which may seem obvious to those of us who are not trying to shake off the effects of an evening of dancing, drinking, and debauchery. As with every other part of this book, we shall omit the mathematical proof for finding derivatives and simply point out that there is a simple process for finding derivatives for first-, second-, third-, fourth-, fifth- or even sixth-degree polynomials, which is described by the following rule: For the function $f(x) = cx^n$, the derivative $f'(x) = dy/dx = cnx^{n-1}$. This rule is quite straightforward when you look at it closely without hyperventilating because it is merely telling us that we can obtain a derivative by reducing the exponent by 1 and then multiplying the original coefficient by the original exponent. So this rule would tell us that the derivative of the function $f(x) = 5x^3$ is equal to $f'(x) = 15x^2$ or that the derivative of the function $f(x) = 2x^6$ is equal to $f'(x) = 12x^5$. In general, the derivative of a polynomial will be determined by its degree and its coefficient. Going back to our rule, we see that x is a first-degree polynomial because its exponent (even though it is customarily omitted) is equal to 1 whereas x^3 is a third-degree polynomial because its exponent is equal to 3 and x^5 is a fifth-degree polynomial because its exponent is equal to 5. So remember that if you need to make an irritating co-worker or neighbor feel inferior during the course of a heated conversation, you can challenge them to differentiate a fifth-degree polynomial in front of a crowd and then carry out the operation without breaking a sweat while your adversary can only shift uncomfortably on his or her feet.

Maxima and Minima

Much of the fun associated with wandering through the treacherous terrain of higher mathematics is the tedium of finding values for a function and then plotting these points on a graph. But there is a certain value to such an activity because it allows one to become acquainted with the information which can be provided by functions, first derivatives, and even second derivatives. Second derivatives? You bet! In the same way that first derivatives tell us about the rate of change of a function, second derivatives similarly tell us about the rate of change of a first derivative. But we are getting ahead of ourselves and need to restrain our enthusiasm or else we might race too far ahead without really knowing where we are headed.

At this point, we need to begin considering how we might use the calculus to carry out useful calculations which can help us to climb the ladder of success. Before we can do this, however, we need to see how the graphical descriptions of various functions can give us important information about these functions and their derivatives. In particular, the derivatives of given functions enable us to determine where the values of a function are minimized or maximized, which leads to many important applications in business and industry.

We can graph almost any function—regardless of whether it is a third-, fifth-, or even eighth-degree polynomial—as long as we have enough patience and paper. You will recall that we merely have to plug in the values for x for any such polynomial to determine the value of the dependent variable y. But this process can get very cumbersome as can be shown by our effort to plot the values of a function having a third-degree polynomial such as $x^3 + 3x^2 + 9$ on a graph. After a few minutes with our trusty calculator, we will find that we have the following values for $f(x) = x^3 + 3x^2 + 9$ as shown in Table 1. This is no doubt a very impressive table for those who are not fortunate enough to see such things in their daily routine but it is nothing more than an ordered arrangement of various values of the function $f(x) = x^3 + 3x^2 + 9$. As such it is no more mysterious than quantum physics or the search for the meaning of life.

Table 1

x	$f(x) = x^3 + 3x^2 + 9$
-3	9
-2	13
-1	11
0	9
1	13
2	29
3	63

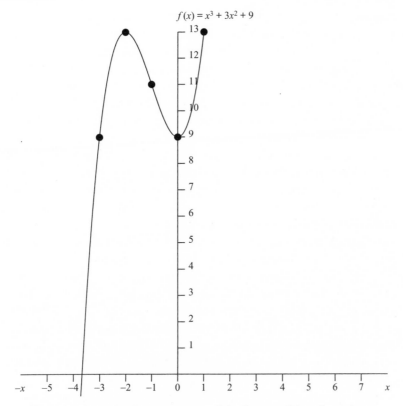

$f(x) = x^3 + 3x^2 + 9$

Figure 16. A graphical representation of the function $f(x) = x^3 + 3x^2 + 9$.

Having obtained five pairs of coordinates, we can plot this function and see that the curve looks something like a snake which has swallowed a live electrical wire, with the curve rising upward from the negative x-axis until it reaches $(-2,13)$ where it then begins to turn downward until it reaches $(0,9)$ at which point the curve then turns steeply upward and roars off into space as shown in Figure 16. This curve has a somewhat serpentine shape but we do not know if our curve is completely accurate because we are only using seven sets of coordinates to plot it. As with everything else in life, we could spend more time on our graph and plot it more accurately by adding additional (x,y) coordinate points but there is a point at which it is simply not worth the additional time and effort to continue refining our graph. Simply put, we will probably not gain great riches or fame by spending the rest of our lives plotting 6,000,000 points describing the function $f(x) = x^3 + 3x^2 + 9$ because there is not very much of a demand for such a service. Indeed, it becomes as pointless as creating ever larger numbers by tacking on ever greater groups of zeros.

But so what? What is the significance of finding where the slope of the function changes direction on a graph? The answer which will have to suffice for now is that the places at which the slope of the curve changes direction from rising (positive) to falling (negative) or vice versa are those points at which the derivative of the function is equal to zero. Of course this means that we have to get our hands dirty again and find the derivative for the function $f(x) = x^3 + 3x^2 + 9$ which, as we saw from our snappy rule outlined above, is $f'(x) = 3x^2 + 6x$. We can then factor this derivative as we are all alone on a Saturday night and our inflatable doll has sprung a leak. These factors are $(3x + 0)(x + 2)$ which we can then set equal to 0 so that we obtain $3x + 0 = 0$ or $x + 2 = 0$ which means that x is equal to 0 or -2. Surprisingly, these two numbers appear to coincide with the points at which the curve of the graphed function $f(x) = x^3 + 3x^2 + 9$ changes direction, as shown in Figure 16.

Life is truly grand when one's calculations seem to have some basis in reality. But we cannot rest on our laurels and we did promise that we would try to make some sense of the relationship between the function and the derivative, which becomes apparent when we overlay the graph of the function $f(x) = x^3 + 3x^2 + 9$ with a graph of the derivative $f'(x) = 3x^2 + 6x$. We find that the two points on the x-axis where the curve of the function changed directions are the same two points at which the graphed curve of the derivative crosses the x-axis. But you need not take our word for it as we can plot out the points of the derivative $f'(x) = 3x^2 + 6x$ as shown in Table 2.

At this point we then create a new headache called the second derivative, which we obtain by taking the derivative of the first derivative $f'(x) = 3x^2 + 6x$ to obtain the second derivative $f''(x) = 6x + 6$. In essence, the second derivative tells us the rate of change of the first derivative. As with the function $f(x) = x^3 + 3x^2 + 9$ and its

Table 2

x	$f'(x) = 3x^2 + 6x$
-3	9
-2	0
-1	-3
0	0
1	9
2	24
3	45

derivative $f'(x) = 3x^2 + 6x$, we find the same values for the second derivative $f''(x) = 6x + 6$, shown in Table 3. What is particularly interesting (and not a mere coincidence) is that the curve of the second derivative $f''(x) = 6x + 6$ crosses the x-axis at exactly the same x-coordinate that the U-shaped curve of the derivative $f'(x) = 3x^2 + 6x$ reaches its lowest point as shown in Figure 17. These graphs of the function $f(x) = x^3 + 3x^2 + 9$ and its derivative and second derivative illustrate the underlying relationships between functions, rates of change (derivatives), and rates of change of rates of change (second derivatives). Although we could wade into a bit of mathematical jargon to explore these relationships, it may suffice to illustrate the ways in which these curves behave in conjunction with each other. In fact, there are some helpful rules of thumb which show how the curves of functions and their derivatives relate to each other. These rules are summarized by Douglas Downing in his book *Calculus the Easy Way*:

Table 3

x	$f''(x) = 6x + 6$
-3	-12
-2	-6
-1	0
0	6
1	12
2	18
3	24

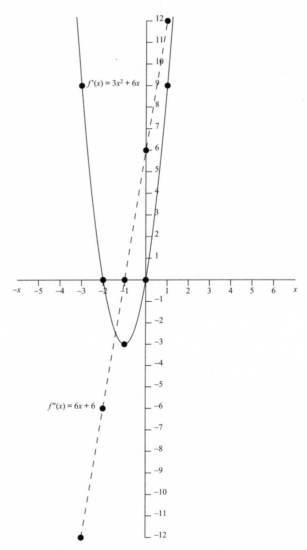

Figure 17. A graphical representation of the derivative $f'(x) = 3x^2 + 6x$ and the second derivative $f''(x) = 6x + 6$ of the function $f(x) = x^3 + 3x^2 + 9$.

1. When the first derivative is positive, the value of the original function is increasing.
2. When the first derivative is negative, the value of the original function is decreasing.
3. When the first derivative is zero, the original curve has a horizontal tangent at that point.
4. When the second derivative is positive, the original curve is concave upward.
5. When the second derivative is negative, the original curve is concave downward.
6. When the second derivative is zero, the original curve has a point of inflection (concave upward on one side of the point, concave down on the other side), provided that the second derivative is positive on one side of the point and negative on the other side of the point.
7. When the first derivative is zero and the second derivative is positive, the point is a local minimum, but when the first derivative is zero and the second derivative is negative, the point is a local maximum.

One can spend hours of fun drawing the curves of functions, derivatives, and second derivatives. The important thing to bear in mind is that all three of these concepts are intimately related to each other. For example, the function can represent the position of our plucky runner Lance and the first derivative can describe Lance's velocity. The second derivative would then represent the rate at which Lance's velocity changes.

Optimal Values

No doubt you have enjoyed our excursion through differential calculus but you may be wondering whether there is any point to mastering the subject other than gaining the satisfaction of knowing that you are intellectually superior to most of your peers. Fortunately, the calculus has a practical side and can be wielded to bring seemingly unsolvable problems to their knees, thereby allowing us to laugh at the heavens and revel in our seeming genius.

One example offered by our friend Douglas Downing relates to the calculation of the optimum sales price of a magazine. Suppose that we are the proud publishers of *Bedwetters Illustrated*, a magazine geared toward providing the latest gossip about incontinent celebrities. Anxious to tap the potentially vast market of adults who crave any news about their favorite diaper-clad television and movie stars, we are concerned that we price the magazine at a rate that will maximize our profits. Profit maximization does not mean that we charge $150 per issue because there are very few people who are willing to pay such a high price for a copy of our magazine, even if they are slavish devotees of the incontinent jet set. The fact that a few souls might be willing to fork over such a hefty sum will not make up for the loss of the many souls who think that such an expensive purchase would be stupid. By the same token, we do not want to price the magazine too low in the hopes of attracting the widest possible audience because we may not be able to make a profit at all, particularly if our magazine revenues are less than the actual cost of producing the magazine. After all, the publication of any magazine, particularly one such as ours which boasts articles featuring complete sentences and half-truths and juxtaposed full-color photographs, is not without its own costs. We will have to pay for paper, for inks, for the machines to print the magazines, and for a building with swank executive offices to house our publication. We will also have to pay the salaries and wages of all the people who will contribute to the magazine—whether they be writers, photographers, editors, or printers—who will periodically go on strike to press their demands for creating a socialist world order. So we will certainly have to quantify both our costs and our revenues to find the optimum price for selling the magazine which will maximize our revenues. Suppose that the number of subscriptions we can sell of our esteemed publication (which is now being offered in a new coffee-table format to attract wealthier buyers) can be expressed using the following formula:

$$n = -7000p + 20,000$$

where n is the number of subscriptions and p is the price per copy.

But we have also determined that it costs us $2.00 to print each copy of our magazine. So we know that we will have to price the magazine somewhere above $2.00 to make any kind of a profit even though we will obviously lose the business of some potential subscribers who might be willing to pay, say, $1.50 per issue. As we would still be losing 50 cents per issue at that price, we would not be particularly disturbed about losing those subscribers as they will obviously not contribute toward the realization of our goal to be the leading gossip photomagazine on incontinent celebrities in the world. However, we will need to figure out the price that maximizes our profits which requires us to define our profit as a function of the price of the magazine. We can define our profit in terms of the following equation:

$$Y = R - C$$

where Y is the profit, R is the revenue, and C is the cost of producing the magazine. Now we know that the cost of each magazine is equal to $2.00n$ and that our revenue is equal to pn (the price of each magazine—assuming a uniform cover price—multiplied by the number of magazines sold). Of course the immediate problem is that our original formula for the number of copies which would be sold has two different variables ($n = -7,000p + 20,000$) which would preclude us from solving for the variable n. But we are well-schooled in the ways of higher mathematics and profit-maximization equations and we can use this relationship between p and n and plug it into the equation for revenue:

revenue $= pn = p(-7,000p + 20,000) = -7000p^2 + 20,000p$

Now that we have proceeded this far, we can put this expression back into the equation which gives us the amount of profit:

$$Y = -7000p^2 + 20,000p - (2.00)(-7000p + 20,000)$$
$$Y = -7000p^2 + 20,000p + (2.00)(7000p) - (2.00)(20,000)$$
$$Y = -7000p^2 + 20,000p + 14,000p - 40,000$$
$$Y = -7000p^2 + 34,000p - 40,000$$

We can take a deep breath as the few remaining flickering neurons in our brains fire enough for us to recognize that we are dealing with a second-degree polynomial. We also know that the derivative of $Y =$

$-7000p^2 + 34{,}000p - 40{,}000$ and can set it equal to 0, which gives us $dY/dp = -14{,}000p + 34{,}000$ or, once set equal to 0: $0 = -14{,}000p + 34{,}000$. This equation can then be solved by shifting $-14{,}000p$ to the other side of the equal sign which gives us $14{,}000p = 34{,}000$ where $p = 34{,}000/14{,}000 = 2.428$. What does this tell us? Some skeptics might say that it tells us we should have pursued a professional career in mid-level management, specializing in the drafting of lengthy ponderous memoranda with amorphous, wishy-washy conclusions. But this is not quite accurate. Quite simply, we are being told that we can maximize our profits (not our sales figures) by selling our magazine for $2.43 per copy.

The distinction between profits and sales is an important one. After all, we could give our magazine away for free and presumably increase our circulation by a factor of 5, 10, or even 15. If we were really interested in getting our magazine in front of the widest possible audience, we could pass out free magazine copies with $5.00 bills taped on the inside. But this approach would not be profitable. Instead we would have to cover the cost of producing the magazine and the additional $5.00 charge per copy in the form of a greenback to entice the general public to pick up the magazine. If we have a bottomless pit of money to spend, then we might not be very concerned with generating revenue. But as we live in the real world in which salaries must be paid and children sent to private schools and liquid assets parked offshore in secret bank accounts to provide some measure of comfort in the event of an unanticipated and costly divorce, we cannot ignore the need to earn a profit. So if we assume that we will be selling our fine magazine to the general public at the bargain rate of $2.43 per copy, we might be interested in finding out the actual amount of our profits.

To determine the total number of magazines we will sell, we need to retrieve the formula $n = -7{,}000p + 20{,}000 = 2{,}990$. Then we calculate the total revenue to be gained from the sale of these issues by retrieving our revenue formula (revenue) $= pn = (2.43)(2990) = \$7265.70$. Now this is not the type of astronomical revenue formula that will inspire multibillion-dollar takeover attempts or stir the imagination with the potential for influencing public opinion to change the course of history because *Bedwetters Illustrated* has not yet become mainstream reading. But every empire must begin with a single brick and so we must be content with our current figures. Having determined the amount of gross revenue our magazine will generate, we then must take into account the cost of putting out each magazine, which leads us to the formula (cost) = ($2.00)(2990) = $5980. In subtracting our cost ($5980) from our revenue ($7265.70), we obtain a profit of $1285.70.

The only question at this point is how we should spend our profit. We could put a down payment on a new corporate jet for the magazine or we might purchase a mountain retreat so that we can conduct high-level strategy meetings with business "acquaintances" without fear of getting caught or we might even plow it back into the company in order to buy better printing presses or a drinking fountain that does not drip sludge. But the important thing to keep in mind about this exercise is that we have used the differential calculus to maximize our profits. The derivative we used in this equation provided us with the point (in this case, the selling price) at which we could maximize our profits. Of course there are those pessimists who would sneer that our profits are still very puny and reflect the anemic interest in bedwetting celebrities. But we also need to bear in mind that we are dealing with an untapped but potentially vast market so that today's modest profits and sales may be only a tiny hint of the glorious fortunes to come. After all, the magazine may represent only the tip of the iceberg of an entire *Bedwetters Illustrated* product line, ranging from sheets that stay dry no matter how long they are immersed in water or other, more disagreeable, liquids to t-shirts which enable wearers to proclaim their incontinence proudly to the world.

Not so fast! You doubt that we have calculated the price at which the most copies of *Bedwetters Illustrated* can be sold? We have been asking you to take our word for it but we should select two other prices—one lower than $2.43 and one higher than $2.43—to see if we are on the right track or merely ranting and raving. So we will try $2.30 and $2.50 and plug them into the formula. Okay, done. We were correct. But you probably want something a little more substantial to justify our conclusion so let us first carry out our calculation using $2.30 as the cover price. The number of magazines we can sell is equal to $n = -7000(2.30) + 20{,}000 = 3900$. By contrast, the higher purchase price of \$2.50 will cause us to sell $n = -7000(2.50) + 20{,}000 = 2500$. So many potential customers (1400 to be exact) will be lost if we raise our cover price from \$2.30 to \$2.50. No doubt many of these consumers will opt to take the money they would have otherwise spent on fine literature and mount a leveraged buyout of a multibillion-dollar conglomerate. But our result does make sense because we would be very surprised to find raising our sales price made more people buy our magazine. There is a more disturbing implication to be drawn from the fact that we are losing more than a third of our subscribers from a very modest 20-cent increase in the cover price; there may be very few people who really want to read about the incontinent side of the entertainment industry. But this is not the place for thoughtful analysis because we are busy trying to verify that we have selected the proper cover price for the magazine. We must now calculate our revenue and cost figures for both examples (cover prices of \$2.30 and \$2.50) and see whether we were in fact correct. We see that the cost of producing 3990 magazines at \$2.00 per copy = \$7980 and that the cost of producing 2500 magazines at \$2.00 per copy = \$5000. Conversely, the revenue side of the equation is fairly straightforward: 3900 magazines at \$2.30 will generate \$8970 and 2500 magazines at \$2.50 will generate \$6250. The profit to be obtained from selling our copies of *Bedwetters Illustrated* for \$2.30 per copy is equal to \$8970 (revenue) − \$7980 (cost) = \$990. By comparison, the profit that will be realized by selling our magazine for \$2.50 per copy is equal to \$6250 (revenue) − \$5000 (cost) = \$1250. Both of these amounts are less than the profit of \$1285.70 which we would obtain by selling our magazine at a price of \$2.43 per copy. Although there are persons who have questioned authority ever since they were first told to wash their hands after playing in the sandbox and who would insist that we calculate the costs and revenues for every other possible price for our magazine ranging from \$0.01 to \$100, we can only respond that the point of the calculus is to simplify our tasks. If we were to graph the profit to be realized at a given cover price (with the profit shown on the y-axis and the price shown on the x-axis), we would obtain a bell-shaped curve that would rise to reflect the increased profits to be realized with an increase in the cover price, with a maximum at the point where the cover price is equal to $2.43, and then decline as the magazine continued to increase in price. We simply do not have the time or the need to continue making incremental calculations of profitability because we are far too busy living the glamorous lives that only those who dabble in the higher mathematics can envision. But our journey into the calculus is only partly completed as we must now consider the integral calculus and the wondrous variety of problems that can be solved with it.

CHAPTER 9

Integral Calculus and Exponential Functions

People used to think that when a thing changes, it must be in a state of change, and that when a thing moves, it is in a state of motion. This is now known to be a mistake.
—BERTRAND RUSSELL

Our previous chapter wandered through the conceptual underpinnings of differential calculus and we were able to become so well acquainted with its basic principles that we can now obtain positions as university professors in higher mathematics. But because hiring committees at most universities insist that their mathematics professors know something about both differential and integral calculus (as well as long division), we must now learn something about this other type of calculus, which has been around in spirit—though not in its technical form—since the time of the early Greeks. The actual concepts of the integral calculus were developed at about the same time as those of differential calculus but the problems which would later become the province of integral calculus were initially examined by ancient thinkers such as the Greek mathematician Archimedes. On the surface, the domains of these two branches of mathematics appear to be wholly dissimilar. Differential calculus is used to analyze rates of change and the slope of the tangent to a curve whereas integral calculus is used to calculate the areas under curves.

241

But they actually complement each other very nicely and, in some ways, can be said to be the reverse of one another.

But we are getting ahead of ourselves. Integral calculus began with early attempts to measure the lengths of curves. Even the dullest among us can appreciate the fact that curves are more difficult to measure than straight lines. Straight lines merely require that one have an accurate measuring stick, reasonably steady hands, and clear vision. Because curves can appear in any of an infinite variety of shapes, the single measuring stick approach is not very helpful. One could amass a massive collection of curved measuring sticks but such an approach would not be very practical. Another approach might be to retrieve a length of string and try to lay it over the curve from endpoint to endpoint and then, after removing the string, measure the length needed to cover that curve. But this is at best an imperfect approach because the string can bunch up and become tangled around our fingers and get wrapped around our neck and choke off our air supply and cause us to suffocate and die. So there is a deep and forbidding danger in mathematical research which is not appreciated by the general public; its members are inclined to think of mathematicians as spending their careers marking arcane symbols on chalkboards and writing obscure, unintelligible papers in dark, dusty offices. Leaving aside these issues, however, we can say that the roots of the integral calculus began with efforts to approximate the lengths of curves.

The most obvious way to measure a curve after the paramedic has unwrapped the string from around our necks and revived us with a tank of oxygen and a morphine drip bag is to divide the curve into a series of smaller segments and then measure each of those smaller segments separately. The idea here is that there is less deviation (and less chance for error) in measuring a series of smaller bits of the curve using the straight line ruler than there is in measuring the single large curve with that same ruler. We might be able to visualize this point if we imagine that we are standing atop a rickety scaffold beneath the dome of St. Peter's Basilica in Vatican City and admiring its curvature. We could measure the diameter of the dome by running from one end of the dome's edge to the other along the scaffold with a measuring tape. But we know that our measured length would be different (and presumably less) than that of the surface of the dome itself from edge to edge. We could try to reduce the discrepancy by dividing the surface of the dome into smaller segments and measuring each segment with our ruler. Because our single straight segment would now be reduced to many smaller segments, they would more closely approximate the diameter of the dome from one edge to the other because we would be measuring these smaller lengths much closer to the actual surface of the dome. We might best understand this point if we imagine that we first began with a single rigid ruler which extended from one edge of the dome to the other. Once we divided the curve of the dome into, say, 50 smaller segments, we might then cut our single ruler into a series of 50 smaller pieces. We would then be able to bring these smaller rulers correspondingly closer to the actual surface of the dome. We would expect to get a more accurate reading of the length of the dome from edge to edge. As we have a great deal of time on our hands, we could divide our 50 segments into 500 smaller segments, each one-tenth the length of the original 50 segments. We could also divide our 50 rulers into 500 smaller rulers. These smaller rulers could be placed against the dome so that there is correspondingly less deviation from the measure of the curve of the dome itself within each segment and its particular ruler.

The very same point can be made if we were trying to measure the inside diameter of a cereal bowl with a ruler. A standard ruler could be laid across the top of the bowl but would be too long and too rigid to be of much use in measuring the curved interior surface of the bowl. If we cut the ruler into one-inch lengths, we can lay them end to end on the inside of the bowl and better approximate the measure of the curve. This will not be a perfect measurement but it will be much closer to the actual value than could be obtained by using our standard 1-foot ruler. We can get an even better result if we slice each inch into quarter-inch pieces and then lay them end to end in the bowl. Our ruler has even more "flexibility" and can more closely approximate the measure of the curve of the bowl. This example is somewhat theoretical because it ignores the fact that we cannot cut the ruler into precise equal segments without some parts of the ruler disappearing as sawdust. Of course even the best of experiments can run aground, particularly if we leave the bowl with the little rulers in it on the breakfast table and a six-year-old uses it for his cornflakes and milk the next day.

But the important thing to bear in mind, regardless of the sogginess of our ruler segments, is that we are following the course pursued by Archimedes and many other great mathematicians in trying to break down curved surfaces into ever-smaller straight-line segments. The problem with such an approach is that we have to deal with smaller and smaller segments and smaller and smaller rulers to carry out our measurements. Fortunately, we can resort to mathematics to carry out approximations in as detailed and as pristine a manner as we may desire. Although our mathematical calculations will not bring with them the satisfaction that can only be gained from stacking soggy bits of rulers end to end in milk-filled cereal bowls, we should be able to calculate the length of any curved surface to any degree of accuracy we may desire.

To return to the St. Peter's dome, we can continue splitting our rulers into smaller and smaller pieces and achieve a more and more accurate approximation of the surface of the dome itself. The supervising cardinal may not be very pleased to see that we have spent the better part of a week taping bits of rulers to the surface of the dome instead of continuing the restoration work for which we had originally been hired to complete—but such is the price of pushing back the shroud of ignorance and advancing the frontiers of human knowledge. If we continue to increase the number of segments, it may be said that the sum of these segments approaches a limit which is the actual length of the curve of the dome.

This all may seem as sudden as a shotgun wedding in the hills of Kentucky but it is actually a very straightforward and intuitive concept. We are using our knowledge of geometry to calculate the rectangular areas (the segments) underneath a nonlinear curve (the dome, the cereal bowl, etc.). Because the area under the curve cannot be calculated as simply as the area of a given rectangle, we must divide the area under the curve into smaller and smaller segments, treating each of those segments as a tiny rectangle as illustrated in Figure 18. We can then calculate the areas of each of those rectangles and add them all together to arrive at an approximate figure

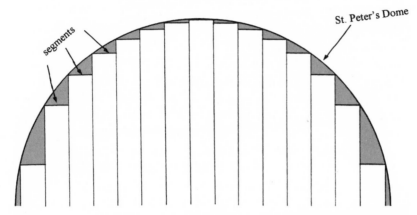

Figure 18. An illustration of the breaking down of the surface of a curve into numerous smaller segments to facilitate the measurement of the area underneath the curve.

for the total area underneath the curve. The more numerous the segments, the more accurate our approximation (as shown by our example of smaller and smaller pieces of rulers) of the area under the curve. At the same time, however, this result is somewhat unsatisfying because we know that there are little bits of area underneath the curve but on top of each of the rectangular segments which we are not including in our calculations as shown by the shaded areas in Figure 18. Now these omissions may not seem like a big deal but such discrepancies drive mathematicians crazy because they are very neat and tidy people who cannot stand to see any type of calculation in which loose ends are involved. Here, our loose ends are all of those little triangular areas perched atop each rectangle under the curve. At the same time, it seems like much ado about nothing as these little sectors are admittedly very small when compared to the entire area under the curve. We suspect with some justification that we could ignore those little triangles entirely and not throw off our calculations by very much. But at the same time we are also troubled by the idea that we could continue this process of dividing the area under the curve into ever-smaller rectangular segments without ever resolving this problem in a completely satisfactory way. What should we do?

Well, we could stand up straight with our shoulders back and our chin high and drop this book in the trash and never give integral calculus another thought but that would be the easy way out. Or we can press on, confident that we will somehow obtain a formula or an equation or even a divine revelation that will enable us to calculate the area under any curve—no matter how convoluted its shape may be.

At this point, it may be instructive to reconsider the function concept which provided such a delightful source of entertainment in the previous chapter on differential calculus. We have a function $y = f(x)$ which connects two points A and B in a two-dimensional (Cartesian) plane consisting of a horizontal x-axis and a vertical y-axis. Let us further suppose that the curve extends over the horizontal axis from the point where $x = A$ to the point where $x = B$. In other words, we have a line describing the function which floats above the horizontal axis, describing the various coordinate positions, which is sandwiched in between the endpoints A and B. The initial point A has an x coordinate (a point on the x-axis) which we describe as a_0. The next point is represented on the x-axis as a_1, the third point as a_2, and so on, so that the last coordinate on the x-axis is marked as a_n, which is also the x-coordinate of point B. Figure 19 provides us with a graphical description of this function.

Now we look at the top of each of the bars that represent our rectangular areas and we see that there are areas underneath the curves (shaded in our diagram) which lie above the rectangular bars. These are the very same areas (in a conceptual sense) as those spaces lying between our ruler and the ceiling of St. Peter's Basilica or our ruler and the inner surface of the cereal bowl. But if we look closely at each of the shaded areas over each of the rectangles, we see that they appear to resemble a certain very extraordinary shape known as a triangle. Now we may be reaching a bit here but it looks like these are a special class of triangles known as hypotenuse triangles which are easily recognized by their right angles. The area contained within each of these triangles does not occupy the entire area above the top of each rectangular bar because that would make our task much too simple, as shown in Figure 20. After all, we would merely have to calculate the area of the rectangle (base × height) and the area of the triangle (½ base × height or, in this case, $\frac{1}{2}\Delta x \times \Delta y$) and add them

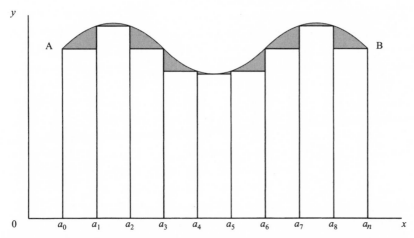

Figure 19. A graphical description of the function $y = f(x)$ between a_0 and a_n on the x-axis.

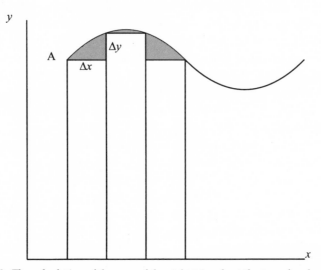

Figure 20. The calculation of the area of the right triangle at the top of each segment underneath the curve makes it possible to approximate the area under the curve itself with greater accuracy.

together and then repeat the exercise for each of the remaining segments and, after adding them all together, obtain a grand total for the entire area under the curve of the function. But life is not so simple because there are still slivers of area lying above the hypotenuse triangles and beneath the curve. At this stage you may be inclined to view these remaining areas as so insignificant as to not justify our concern. But when you are a mathematician, you cannot be so cavalier about little spaces tucked here and there beneath your function because you are in a field which demands precision—not "sort of" completeness. Even though we know that the base of each of these triangles is equal to Δx and the altitude of each of these triangles is equal to Δy, our ability to calculate the areas of each of these triangles will only get us part of the way to our goal of being able to calculate the area underneath the curve. As we allow Δx to go to 0 (which happens as we add more and more segments between points A and B or, alternatively stated, increase the number of intervals along the horizontal axis between a_0 and a_n), the sum of the hypotenuses of the right triangles approaches a limit which is the length of the curve itself. In a sense, we are lending an additional degree of refinement to our method of approximation because we are using both the rectangles which figured in our measurement of the dome of St. Peter's and the inside of the cereal bowl along with right triangles to measure most of the remaining area between the top of the rectangle and the curve itself. As with the differential calculus, we are allowing the number of segments to increase, thus causing each additional incremental change along the horizontal and vertical axes to become smaller and smaller. The smaller the incremental differences, the more closely our calculations of area can approximate the area underneath the curve.

But might there be a better way to approximate the areas under a curve? Up to now, we have focused our energies on calculating these areas by carving out successively smaller spaces under the curve. This is a very rewarding activity in and of itself because it occupies our time and keeps the more violently inclined of us inside our homes perched over our desks and away from other people who might otherwise be harmed. But we have seen that there is a limit to the accuracy of this use of rectangles and triangles to approximate the area underneath a curve. We live in the space age and we do not want to be using Model-T mathematics if there is something more sleek and sophisticated with which we can tackle our mathematical problems. Our discussion of the differential calculus led us to the concept of the derivative; it seems only fair that we should have an equally "cool" mathematical device to assist us in our exploration of the integral calculus. It might help to figure out the point of dealing with the integral calculus in the first place.

The differential calculus is very helpful in allowing us to begin with a position function and end up with a velocity function. In the example in which Galileo was tossing objects such as lawn furniture off the Leaning Tower of Pisa, we began with a position function which expressed the altitude above the ground from which the debris was dropped. By differentiating the position function, we obtained a velocity function which showed us the rate of speed at which the object was falling to earth. When we deal with problems relating to integral calculus, we differentiate backwards in that we begin with a velocity function and then determine a position function. No doubt this news will gladden even the coldest of hearts because many people often complain of their inability to sleep soundly at night due to their frustration at not being able to derive position functions from velocity functions. Such is the wonder of the age in which we live and so we will joyously go forward and explore some of the conceptual underpinnings of the integral calculus which we could view as a sort of fraternal twin of the differential calculus.

The best way to get a handle on the integral calculus is to consider a problem in which we need to determine the initial position of an object when we know its velocity function. But we should first remember that in the differential calculus we used the process of differentiation to get us from the position function to the velocity function. The reversal of this process is not so difficult as it may first seem because we are essentially carrying out the procedures of the differential calculus in reverse. Ooh, this does sound like fun! So we first need to make sure we have a grasp of the reverse differentiation process because our previous venture into differentiation was a little more complicated than a simple visit to the playground. If our original (position) function is $3x^3 + 2x + 2$, then we know from our previous discussion that our derivative (velocity function) is equal to $9x^2 + 2$. Conversely, if our initial velocity function is equal to $9x^2 + 2$, then we can reverse the differentiation process and obtain a position function equal to $3x^3 + 2x + 2$. The reverse differentiation leads us to what we would be tempted to call an antiderivative (the evil twin of the derivative) but which mathematicians who were not so familiar with the dark forces in comic books decided to call the "integral."

Now let us move ahead and consider how we might actually make some use of the integral function and, indeed, integral calculus as a whole, in the real world. Suppose that we have hired Captain Bob to take us out on his boat for a day of fishing. We know nothing of Captain Bob's extreme susceptibility to seasickness or his hydrophobia (fear of water) or his chronic amnesia or his agoraphobia (fear of open spaces) or his inability to use a navigational map or even his ineptitude with a compass but we do know from his telephone book ad that he is a "sailor's sailor" who "loves the life of the sea!" Because of his amnesia, however, Captain Bob is unable to tell us the location of the dock from which he will be leaving to begin our fishing trip. (Unfortunately, Captain Bob forgot to have the address of the dock put into the phone book or to give it to directory assistance.) But because Captain Bob has the speed of his ship tattooed to his hand, he is able to tell us that the speed of his boat is equal to $v = 6t$ where v is velocity and t is time. At least he thinks it is the speed of his boat or perhaps a verbal signal which can be used in a tight poker game or even the initials of a former girlfriend. But because Captain Bob is the least expensive captain listed in the telephone book, we decide to take a chance and to go ahead with our trip. But after we hang up the phone, we decide that we need to determine where we can find Captain Bob. This is where our knowledge of the integral calculus can come in handy. But it is important to formulate a plan of attack so that we will not be wasting our time. First, we look at a local map and quickly eliminate any landlocked locations from consideration because they lack water access. This leaves us with a stretch of coastline indented with a harbor and the mouth of a river. Because there are many marinas in the harbor, we call Captain Bob back and ask him to be more specific about his location. He still cannot recall his address but he knows that he is within one-tenth of a mile of the "big tall tower" which we know to be the famous Zoots Tower, which rises 500 feet above the surrounding land and provides diners at the world-renowned Zoots Sausage Palace (which is perched precariously at the top of the tower) with troughs of sausage links swimming in a soup of lard.

What do we have to work with? We have our velocity function $v = 6t$ and we have a point of origin which we are assuming is 500 feet from the dock at which Captain Bob will be guiding his beloved ship "Bounty" out to sea, laden with passengers and a unique fudge chum which Captain Bob swears will draw any fish with a hankering for chocolate. Coincidentally, the Zoots Tower is several hundred feet inland from the docks so we will have to trust that Captain Bob's statement has some basis in reality. But we will run into a problem when we try to arrive at a position function by integrating the velocity function $v = 6t$; we will see that it is, by itself, too indefinite for us to draw any definitive conclusions about the original position of the ship. Yet our point about its lack of precision will have to wait for a moment because we first need to integrate the velocity function to obtain the position function. If we are beginning with a velocity function $v = 6t$, then by reversing the normal differentiation process, we know that we will end up with $3t^2$. The only number which, when multiplied by 2, is equal to 6, is 3. Therefore, we know that the original position function (when the velocity function is $v = 6t$) is equal to $3t^2$. Or do we?

The only problem with relying solely on this integration procedure is that we do not know whether the original position function is equal to $3t^2$ or $3t^2 + 8$ or $3t^2 + 74$ or $3t^2 + 516$. This is what we mean when we say that the function is indefinite. We do not know how far from the point of origin (the Zoots Tower) the boat actually begins its journey (whether it is 0 feet or 8 feet or 74 feet or 516 feet). We can probably narrow down the choices somewhat because we know from our review of the map that the "Bounty" must be at least several hundred feet away from that landmark (unless it happens to be a special amphibious ship which can travel down city streets as well as it can navigate coastal waters) because we know that the Zoots Tower sits back from the shore. But we cannot be more specific if we have no other information.

What do we do? Being people who dare to dream big dreams, we could decide to scrap the fishing trip altogether so that we could go spend our time in some of the local watering holes instead of grappling with integral calculus. But we did not take all of those remedial classes in elementary mathematics only to allow ourselves to be cowed by a bothersome gnat of a concept known as the integral calculus. So we need to develop some rules to enable us to determine the exact integral because we cannot rely on random guesses for our calculations. The advantage of random guessing is that we do not have to clutter our minds with tedious rules and can thereby allow our imaginations to run wild. But the disadvantage is that the absence of these rules makes it almost impossible for us to determine the correct answer in a short time. If we are going to leave ourselves with any time to visit some of the local watering holes and master integral calculus, then we do need to devise these rules. Fortunately, this work has already been done for us by people who were much more diligent than we were in turning in their mathematics assignments. These persons (who were always unpopular in school because their test scores would shatter the grading curve used by the teacher and cause the rest of us to appear like morons) saw that the position function's indefiniteness meant that it could be described in any number of ways. In other words, we could take the relationship $dx/dt = 6t$, $x = 3t^2 + C$ (where C is any constant number) and graph any number of curves, all of which would appear to be exact copies of the curve shown in Figure 21. The only difference would be that the curves would be shifted upward by the amount equal to C. In other words, the curve shown reflects a function in which $C = 0$. If we make $C = 2$, then the entire curve shifts upward by two units. In all other respects, this curve is indistinguishable from the previous curve in which $C = 0$; the only difference is the vertical displacement of the curve to its new position.

No doubt you are experiencing a warm feeling of accomplishment at this breakthrough. We are not yet at the point where we can formulate our rules because we have merely provided a graphical description of the ambiguity which is preventing us from determining our position function. At the same time, however, we do know that there is no function other than $v = 6t$ with a derivative having a curve identical to that found in Figure 21. So it is somewhat mislead-

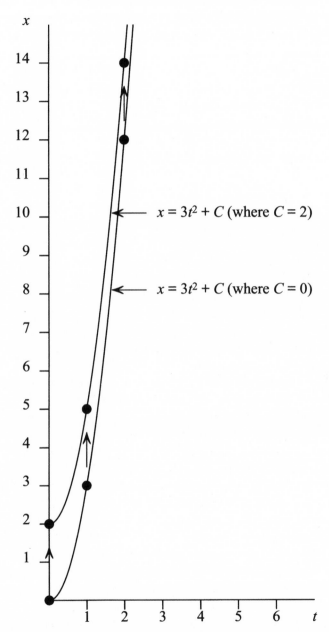

Figure 21. An illustration of two identical functions which differ only in the values given to their constants.

ing for us to speak of this integral as being indefinite because it is only indefinite in that we do not know how far upward the curve lies on the graph. In other words, we do not know the value of the constant C or, to relate it back to our question as to Captain Bob's initial starting position, the distance from the point of origin (the Zoots Tower) at which Captain Bob begins his journey from the docks.

While we do have a problem of indefiniteness, it is not so hopeless as it may first appear. The curves differ only by the vertical displacement. Each curve has the same shape so we know that a function which is described by a different shape will not work because its derivative is not equal to $6t$. The only curves which are appropriate are those that are similar in shape to those in Figure 21. The problem is to determine which of these curves is the correct position function so that we can determine where Captain Bob is at any given moment. Now Captain Bob's wife has no interest in determining Captain Bob's location at any given moment but that is due more to her having tired of Captain Bob's fishy smell than any aversion to integral calculus. But we are interested in determining Captain Bob's location so that we can carry on with our fishing trip. So how should we proceed?

We could stare at all of the curves and begin uttering Gregorian chants in the hopes that we might experience a divinely inspired revelation which would enable us to pick out the one special curve that will allow us to track down Captain Bob. Unfortunately, chanting does not typically lead to greater clarity of thought in any branch of mathematics including calculus. Success in this endeavor requires that we understand the significance of the discrepancies between each of these curves. If we cross our fingers, make a wish upon a star, and do a quick mathematical trick, we will see that the only difference between the curves is the position where Captain Bob begins his journey. For a position curve in which the position was equal to 0 and the time was equal to 0, for example, we would be saying that Captain Bob was sailing from the Zoots Tower at the beginning of his journey. Because the Zoots Tower is several hundred yards from the shoreline, this would not appear to be the appropriate choice because even Captain Bob does not usually steer his boat down busy city streets. So we have to find a position function in which the position is equal to at least the distance from the Zoots Tower to the shoreline. If the Zoots Tower is 500 feet from the shoreline, then we will want to pick a position function with a constant value of C which is at least equal to 500 feet. So if Captain Bob's ship starts 500 feet from the Zoots Tower (the point of origin), this means that our x value = 500 when our time value $t = 0$. This will at least enable us to tell which of the potentially unlimited number of curves scattered up and down the vertical (position) axis of Figure 21 is the correct one. This will also tell us Captain Bob's position function and enable us to find Captain Bob. So we need to solve for the constant C for our position function $x = 3t^2 + C$. If our time t is equal to 0 and our value C is equal to 500, then we get the following solution and our indefinite equation becomes specific: $500 = 3(0)^2 + C$. This equation tells us that C must be equal to 500 when Captain Bob's time t is equal to 0. As a result we obtain a position function for Captain Bob which is described with the following equation: $x = 3t^2 + 500$.

You see the problem with this approach is that one cannot simply rely on calculating the integral to determine the position. We must also have some knowledge of Captain Bob's position at the start of his journey. Here we are saying that Captain Bob begins his journey 500 feet from the point of origin (the Zoots Tower). Once we have established a starting point, then we can determine Captain Bob's position (assuming he continues to maintain a constant velocity) so that we can locate Captain Bob at any point after he has left the dock. Although we had set out to find Captain Bob's original position using the integral calculus, we have since found that we need to know his original position to be able to determine his position anywhere along the position function. Once we find the initial position of Captain Bob's boat relative to our point of origin (Zoots Tower), then we can insert that quantity in place of the variable C (constant) and then solve for the indefinite integral. At that point, the angels will sweep down from the heavens, the grounds will open and swallow up all evil persons, and maybe, just maybe, the Chicago Cubs will win the World Series. On a more mundane level, we will then be able to specify the position function and find Captain Bob at any point in his journey so long as his velocity function is equal to $6t$. True, we do need to be able to find Captain Bob's starting point. But once we have located the dock from where he begins, we can take a chase boat if Captain Bob forgets to wait for his passengers and simply sets off with an empty boat.

So $x = 3t^2 + 500$ is our only guide to tracking down our wayward skipper. Assuming that Captain Bob is already far out at sea, barreling ahead with smoke billowing out of the engines, we only need to know at what time Captain Bob began his journey to determine how far out into the sea he is located. With many situations, it might be difficult to determine the actual time at which the voyage began. But Captain Bob has a steel plate in his head which picks up the ten o'clock chimes from the nearby Flaming Blade Belltower. Because of a post-hypnotic suggestion placed in Captain Bob's subconscious some years ago at a party, Captain Bob is compelled to begin his voyage when he hears the ten o'clock chimes. Although this suggestion has guaranteed that Captain Bob is always very punctual in his morning departures, it also means that he has no choice but to head out of the docks every evening at ten o'clock and float about for four or five hours in the darkness until returning back to shore. For our purposes, we know that if we get to the dock at 2 seconds past 10:00 P.M., then we can figure out that Captain Bob's position is $3t^2 + 500$ or $3(2)^2 + 500$ feet or 512 feet from the point of origin (the Zoots Tower). As the dock is 500 feet away from the Zoots Tower, we know that we can ignore the constant and determine that Captain Bob's boat is 12 feet from the dock. After 10 seconds have elapsed, we will know that the boat will be $3(10)^2$ or 300 feet from the dock. If we add the constant C back into the position function, we will see that the ship is also 800 feet from the Zoots Tower. Unless we are particularly concerned about knowing Captain Bob's proximity to the Zoots Tower, however, we will probably find it easier to disregard the constant and focus merely on the boat's distance from the dock. But we do need to point out that the position function does not extend indefinitely far into the future. Captain Bob's boat will not continue out to sea for months on end nor will the boat's speed continue to increase without limit. But this example does illustrate how we can use the most basic feature of the integral calculus to derive a position function from a known velocity function. We should point out that the nomenclature for describing an integral (the so-called antiderivative) is the reverse of that which is used to describe the derivative of a function in the differential calculus. Mathematicians typically use the symbol $F(x)$ to refer to the integral of a function in the integral calculus. This is not altogether surprising because the integral is a backward derivative.

Now we can press ahead and fully immerse ourselves in the graphical description of the integral calculus which is certainly more enjoyable than having bamboo stuffed under one's fingernails or having one's toenails plucked out by a sadistic pedicurist. As with the differential calculus, there is some effort involved in learning the basic concepts of the subject. But we can begin by discussing the problem of finding the area underneath a specific segment of a curve. If we take a look at the curve shown in Figure 22, we see that the curve describes a function $y = f(x)$. This curve represents the upper boundary of the shaded area underneath the curve; it is bounded to the left and the right by lines which pass through the horizontal axis through points A and B (parallel to the y-axis) and intersect the curve. Now we can draw any number of additional lines between A and B which also rise upward from the horizontal axis and intersect the curve. Each of the segments along the horizontal axis between any of two adjacent lines may be referred to as a subinterval of the horizontal axis. But you will be pleased to know that we can take any one of these subinterval lines up to the point where it intersects the curve

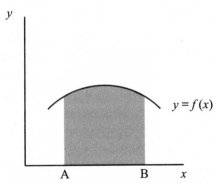

Figure 22. A graph of the function $y = f(x)$.

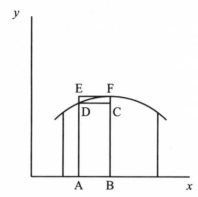

Figure 23. An illustration of the basic method for approximating an area under a curve by drawing ever smaller segments such as ABCD and ever smaller triangles such as DCF.

and do something quite unimpressive: we can draw two rectangles— one of which is completely under the curve and one through which the curve cuts as shown in Figure 23. Although one could stand in awe of the beautiful colors used in this monochromatic drawing or admire the magnificent brushwork, we are more interested in examining the ramifications of the graph itself.

We can see that there are two rectangles in Figure 23, one of which (*ABCD*) is completely contained within the larger rectangle (*ABFE*). The curve cuts through the smaller rectangle *CDEF*. We can then approximate the area of the interval under the curve by calculating the area of the *ABCD* rectangle which, as every tenured professor of mathematics knows, can be done merely by multiplying the width of the base of the rectangle (along the horizontal axis) by its height. We can then calculate the approximate area contained within the triangle *CDF* (1/2 the base multiplied by its height). Once we add the area of this triangle to that of the rectangle, then we will have attained true wisdom and be at peace with the world. We will have also approximated the area underneath the curve for the subinterval between the two points *A* and *B*. Of course we can be as precise as we like in these approximations and create smaller and more numerous subintervals. Indeed, we can create thousands, tens of thousands, or even hundreds of thousands of subintervals so long as we do not mind giving up an active social life and being called a "geek" by complete strangers on the streets. But the important point is that if we continue this process over and over again, we can approximate the entire area under the curve. To use our snobby mathematical lingo, we can say that as we continue to increase the number of subintervals on the horizontal axis, the sum of the areas contained within the subintervals more closely approximates the area under the curve. This sum of the many rectangles and triangles under the curve is called the *definite integral* of the function $y = f(x)$ between the values of A and B (assuming those are the boundaries described on the horizontal axis which enclose the opposite ends of the area underneath the curve). Our good friend and mathematician Leibniz devised a shorthand expression which is still used today to describe this definite integral:

$$\int f(x)\, dx$$

Knowledge of the purpose of this integral brings with it real power. You can interrupt any conversation about any topic and scribble this expression for the definite integral on a napkin or a piece of paper or a chalkboard or even a wall and you will invariably impress people. Whether you have the slightest clue as to what you are talking about is of minor consequence. Most people who have passed through the gates of an accredited university have seen the definite integral; comparatively few have any idea of or clearly remember its relevance to mathematics. But knowledge is power and the best kind of knowledge is that which takes advantage of the ignorance of other people who could not be bothered to study their calculus lessons when they decided instead to attend ritual animal sacrifices while in school. So you should not be reluctant during a sales meeting to doodle a definite integral after your colleagues have discussed whether your company should focus its marketing efforts on pleated skirts or slacks and declare that your analysis of the definite integral leaves you unable to draw any conclusion but that the company should sell whatever will make you the largest commissions. Of course this is not the proper use for the definite integral but it is a use that can be beneficial to you in your career so long as none of your colleagues bothered to attend calculus class. If you work around a bunch of mathematical types, however, this will not be a good strategy as everyone will soon know that you have no idea what you are talking about.

So how can we use the definite integral and really impress our colleagues? We know that we can describe each of the subintervals along the x-axis in terms of changes in the points along that axis or Δx. Each of the bases of these subintervals has a separate value for Δx which may or may not be equal. The altitude of the rectangle is found by extending a line parallel to the y-axis from the x-axis to a point intersecting the curve. The area of each of these rectangles is equal to $f(x)$ multiplied by Δx (the change in position along the x-axis). The sum of all the values for Δx is the sum of the areas of the subintervals under the curve. As we increase the number of subintervals without limit and the value for Δx approaches 0, the limiting area is written as $\int f(x)\,dx$ where dx represents Δx.

From our discussion thus far we would assume that the definite integral is an area. Yet it also is a number and a sum of numbers. As pointed out by Kasner and Newman, "one of the achievements of the integral calculus has been the determination of the moment of inertia of solids"* which in turn has made it possible for engineers to build bridges and dams, most of which do not fall down or collapse. Integration makes it possible to consider the gigantic pressures exerted by water against a dam by determining the water pressure at an arbitrary point and summing it over the whole face of the dam (which is analogous to the area underneath the curve) so that the total force can be determined. The integral calculus can thus be used to identify the center of gravity of any plane or solid figure which applies to the particular function which describes that figure.† Engineers are only one group of persons who have benefitted from the many applications of the integral calculus. It has also helped to provide employment opportunities for numerous mathematics teachers and the companies that publish mathematics textbooks laden with intricate graphs, equations, and drawings which sell for as much as a naval destroyer or a fighter plane.

---

*Edward Kasner and James R. Newman, *Mathematics and the Imagination*. New York: Simon & Schuster, 1940, p. 340.
†Ibid.

The concept of the definite integral is nicely complemented by that of the *indefinite integral*, which is arguably of even greater value to mathematicians because it underscores the relationship between the derivative and the integral. "The definite integral of the function $y = f(x)$ is a number determined by an interval of definite length and a portion of the curve $y = f(x)$ defined over that interval. When the interval is extended from a fixed point to a succession of others, to each of these there corresponds a value of the definite integral. This correspondence, this function, is the indefinite integral of the original function $y = f(x)$ and is symbolized by $\int f(x)dx$."* You may be wondering whether there is a point that is being made in this fancy mathematical chatter. If you look at this mathematical chatter closely, however, you will see that differentiation and integration are inverse operations of each other. You may have already suspected this to be true when we were showing that differentiation of a position function gives us a velocity function and integration of a velocity function gives us a position function. This realization is truly "way cool." In a sense, differential calculus is the reverse image of integral calculus— much in the same way that addition is the reverse image of subtraction and multiplication is the reverse image of division. "Starting with the function $y = f(x)$, upon differentiating we obtain dy/dx. What do we get upon integrating the function dy/dx? The motif of the calculus is hereby revealed, for we obtain the original function, $y = f(x)$. The indefinite integral of the function $y = f(x)$ is another function of x which we shall denote by $y = F(x)$. Of course, the derivative of $y = F(x)$ is $f(x)$. Every function may thus be regarded as the derivative of its integral and as the integral of its derivative."†

Exponential Functions

Having wandered through the integral calculus, it now remains for us to shift gears and consider a special type of function called an exponential function. What makes this function so interesting is that it sounds very esoteric but it is actually a mathematical function

---

*Ibid., p. 341.
†Ibid.

which appears in a variety of real-world situations. One of the most common examples of an exponential function can be illustrated by graphing a population's growth over time. In other words, the rate at which the population grows over time is plotted on a standard x–y diagram with the population itself marked along the vertical axis and the time along the horizontal axis. We have all seen these graphs which begin near the bottom of the vertical axis and bump along the horizontal axis until reaching the modern era at which time—depending on the scale of measurement—the graph heads upward in an increasingly vertical slope. So even if we began at the time 4004 B.C. when, as every evolutionist knows, Adam and Eve were living in the Garden of Eden and plot the growth of the human population, we will see that it does not grow very rapidly in the beginning even though Adam and Eve may have had many children who did not have terribly strong feelings against incest. Because we are starting from a point at which there are only two people, there is not very much of a population to take into account. At the same time, living conditions at that time were very primitive and the mortality rate was high. Because it was difficult to find food, people fought with each other and dropped big rocks on each other's heads under cover of darkness. So life was, to quote Thomas Hobbes, "nasty, brutish, and short." As a result, the population did not grow very fast in absolute terms, even though the gene pool probably became overrun with recessive genes.

But living conditions improved with time as humanity moved from a hunting and gathering society in which people relied for their food on animals which happened to be passing by their villages or on the fruits that could be found in the wilderness to a farming society in which livestock was domesticated and crops harvested. The domestication of the food supply enabled humanity to break away from the bare subsistence living to which it had previously been accustomed and finally grow its own food in large quantities. Freed at last of the need to chase wild animals day in and day out, humans now found themselves with greater leisure time which they could spend engaged in backbreaking toil in the fields. But the availability of a more reliable food supply also contributed to a substantial increase in the rate of population growth. So the curve of the graph showing human population growth over time began to turn upward appreciably as the centuries passed. As humanity passed through the medieval era, the Renaissance, and the Industrial Revolution, the rate of the world's population growth continued to increase. In 1650, the world's population passed 500 million persons and by 1850 it had reached 1 billion persons. By 1975, there were more than 4 billion persons on the planet and this number was expected to increase to 6 billion by the end of the millennium. Clearly, the rate of cuddling and caressing and (dare we say) intimate relations has kicked into hyperdrive in the past century. The corresponding curve shows that greater unit increases in the world's population (where 1 billion is a unit) continue to occur in correspondingly shorter amounts of time. Whereas it took many thousands of years for the world's population to reach 500 million persons, it took only 200 years for it to reach 1 billion persons and 125 years for it to exceed 4 billion persons. If this information is plotted on a graph (see Figure 24), the curve seems to resemble the trajectory of a rocket being fired into outer space. Even more horrifying is the prospect that this curve will continue upward without limit so that the population begins to double at an ever-increasing rate. Given this rapid rate of growth, our distant descendants might someday find

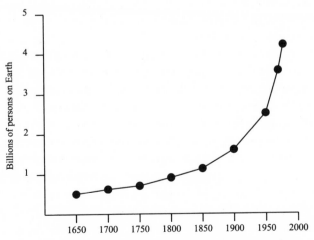

Figure 24. The growth of the world's population over time.

themselves living in a world which has 100 billion people who are packed together like passengers on a subway train. The only problem here is that the entire world would be the subway train and people would always be elbow to elbow with each other. There are some "touchy feely" types who would like to live in a world in which everybody's sweat-drenched body was constantly rubbing against his or her neighbors. For most persons, however, this would be a very dismal life indeed. Fortunately, this population growth would not continue indefinitely because the rate of population growth would eventually outstrip the food supply and there would be widespread famine and pestilence and disease and death. So we would eventually return to a happier time, after burying nine-tenths of the population, in which people were not elbow to elbow and face to face. But why should this be the case? Why would we have to have such a catastrophe? Is it not possible that improved technology in farming and animal husbandry would enable us to continue producing greater amounts of food for a steadily swelling population which continued to increase without limit? Well, there is a certain truth to this idea which has been borne out in the modern era as increasingly efficient food production techniques have made it possible for a steadily dwindling number of farmers to feed greater and greater numbers of people. But farming relies on resources other than technology such as water and soil. As the population increases, correspondingly greater demands are made on surrounding lands, including farmlands, for housing, businesses, roads, and yo-yo factories. At some point, the reduction in the amount of arable land available for farming will begin to constrict the ability of the technological improvements to continue increasing food supplies. The result could be unimaginable death with rotting bodies lying all over nicely landscaped lawns. This gloomy scenario was predicted by the British social philosopher Robert Malthus, who was not known to be the life of any social gathering. Malthus had based his conclusion on the fact that food production increases incrementally (e.g., 1, 2, 3 units) with more intensified cultivation and the addition of new lands, whereas populations increase geometrically (e.g., 2, 4, 8, 16, etc.). Malthus did not apparently take into account the possibility that humans might use birth control but he certainly made clear what he viewed as dismal prospects for humanity's long-term fortunes.

This learned digression on populations leads us to a very famous irrational number which is represented by the letter e. The number e, like the number π, is a sum of an infinite series. The derivation of this number is not important for our discussion but it can be expressed by the following equation: $e = 1 + 1/1! + 1/2! + 1/3!...$ and so on without end.* If we sum all of these fractions together, we find that the value for e is equal to approximately 2.7183. No doubt the feeling of excitement upon learning the decimal value of e rivals that which you felt when you learned that the decimal value of π is equal to 3.1416. But the important point is that we can graph the function $f(x) = e^x$ which is a *continuous function* and can be used to express rates of exponential growth. This is not as useless as it may first seem because exponential functions are important to many aspects of modern life including banking, demographics, and science.

Now that this topic is beginning to catch your fancy, we can examine these exponential functions in a little more detail. As we are all interested in money and finance and accumulating as much stuff as we can before we drop dead, we must familiarize ourselves with the phenomenon of compound interest. Any chief economic officer of a major bank knows that there is a handy formula for calculating the amount of interest earned over a given period of time. Let us assume we want to invest a certain amount of money—P dollars—in our bank. If P dollars is invested at an interest rate of r per cent for a period of 1 year, then the total amount of money in the account at the end of that year will be equal to $P + P(r/100)$. If we invest $100 at an interest rate of 10 percent and we leave this vast fortune in the bank for the entire year, then we will have a total of $110 after one year. But what if we see the possibilities of amassing huge wealth and leave the money in for a second year? Then we will modify our formula as follows: $P(1 + r/100)^2$. At the end of 2 years, we will have a grand total of $121. Unfortunately, this example posits a happy world in which the Internal Revenue Service does not exist; our rate of return would be correspondingly lower if we took into account the taxes on the interest which would be imposed at the end of each calendar year.

---

*The symbol "!" indicates that the number is a factorial so that 3! is equal to $3 \times 2 \times 1$ and 5! is equal to $5 \times 4 \times 3 \times 2 \times 1$. Hence 1/1! is equal to $1/1 = 1$ and 1/2! is equal to $1/2 \times 1 = 1/2$.

But we are here to teach principles and cannot become mired in the messy details of everyday life. As a result, we will pretend that our money will continue compounding without being siphoned off in part by rapacious governmental authorities. Every time we add interest to the principal P, we augment our wealth. If we were to allow our account to continue compounding for 500 years, then we would end up with a very large sum which, unfortunately, would do us very little good as we would have been dead for more than four of those centuries. But if we placed our money on deposit with a very sturdy, stuffy bank and managed to keep the account going for five centuries at 10 percent per year, our original $100 investment, based upon our formula $P(1 + r/100)^{500}$ where 500 is the number of years in which the interest is being allowed to compound, would be equal to $49,700,000,000,000,000,000,000,000! This unimaginably vast sum is equal to 49 septillion dollars, which is a sum roughly 5 billion times greater than the world's current economic production. You would certainly be quite the impressive person if you could buy and sell entire countries or even continents by writing a check. Of course there would be a certain amount of frustration with the fact that one-five billionth of your wealth would be enough to own everything in the world. The remaining part of your fortune would cease to have any real meaning because there would be nothing else that you could do with your money. Actually, the only thing you could do would be to flood the world's money supply with vast quantities of dollars, marks, rubles, yen, and any other currency you happened to own. The ultimate result is that we would have a world in which a loaf of bread would cost some $5 billion and a $100,000 home would cost $500 trillion. Such a price tag would be very steep, even for a two-income professional couple. However, the lesson of compound interest is not lost on us in that it makes it possible to build vast fortunes from very meager beginnings and ranks second only to marrying into a very rich family as the best way to become a self-made tycoon.

But the $49 septillion is the amount that we would have after 500 years if our money earned a steady 10 percent a year compounded annually. If we alter the interest rates, we can see how the amounts change dramatically. Suppose that we put our $100 in an account which earns only 5 percent per year compounded annually. We will find that at the end of 500 years our ungrateful descendants will have the comparatively meager sum of $3,930,000,000,000! Even allowing for inflation, $3.93 trillion is nothing to sneer at but we will probably find that these distant relatives will not even bother to make a toast to us on our birthdays or even come by to leave flowers at our gravestones. Similarly, if we were to invest that $100 in an account paying 8 percent interest per year, then our undeserving relatives would get to split the princely sum of $5,150,000,000,000,000,000! Even the most unpleasant people can smile upon learning that they will be sharing in a fortune of some $5 quintillion. But there will always be that greedy, selfish member who would point out that had our $100 been loaned to an individual at a rate of 21 percent and that individual and his heirs and his heirs' heirs failed to pay the loan back for 500 years, then the total amount would be equal to $24,700,000,000,000,000,000,000,000,000,000,000,000,000,000! Of course this very same person would overlook the fact that no debtor could have amassed the amount of money necessary to repay this debt. But that unfortunate circumstance would merely give the relative another reason to grumble about the failure of people to pay their debts when due.

Because it is not very realistic to think that any account will be left alone for 500 years, we might try a period of 100 years, particularly when there are more and more people living beyond the century mark. If we take that same $100 and invest it in an account paying 5 percent interest for 100 years, then it will be worth $13,150.13. If the account pays 8 percent interest, then we will collect $219,976.13. But if we are fortunate to have our money in a well-managed account which earns 10 percent each year, then we will earn $1,378,061.23 on the date of the account's one hundredth anniversary. And if we are especially fortunate to find a very stupid bank which pays 21 percent interest for its deposits, then we would theoretically (assuming the bank did not go out of business) end up with $19,000,000,000! Now we are talking about some real money. Unfortunately, this bank would not continue paying such a high rate in a stable country in which the rate at which it could loan funds ranged between 9 to 12 percent—unless, of course, the bank was very stupid and very big.

Parking money into bank accounts and forgetting about them for 100 or even 500 years seems to be quite a clever way to amass wealth. But why are there not more millionaires around who had the fortune to have such thoughtful ancestors? Because people do not customarily think of the distant descendants 15 generations in the future. After all, no person reading this book has probably ever said to himself "I should put some money in the bank for my great-great-great-great-great-great-great-great-great-great-great-great-great-great-great grandchildren." Even if they have engaged in such farsighted planning, there is always the danger that some intermediate heir such as a grandson or a great-great granddaughter will insist that they be given the funds as part of the settlement of their respective parent's estate. In short, it is very difficult to keep a bank account out of the hands of grasping heirs who are not at all interested in your efforts to better the lives of your 15th-generation descendants.

So we see that if we compound our interest once a year, this formula will do quite nicely for calculating the amount in our account at any given year. But what if we bank with an institution which compounds our funds twice a year or four times a year or even more? Then we need to modify our basic compound interest formula as follows: The total amount of money M we will have at the end of t years when we compound the interest n times per year is given by the following formula:

$$M = P(1 + r/100n)^{nt}$$

As the frequency of the compounding increases in a given year, whether it is 4, 8, 12, 20, or 50 times, the calculation becomes more and more cumbersome. Imagine how much paper and how many pencils we would have to use if we allowed the rate of compounding to increase without limit! This would mean that we would be compounding our interest continuously. Now the idea of compounding our interest without limit seems like an obvious gateway to riches until we realize that we are compounding ever smaller amounts as we allow the rate of compounding to increase without limit. What is curious is that the continuous compounding of interest can be expressed as being equal to the principal P^e raised to the exponent

$rt/100$. So the amount of money we obtain by continually compounding our interest can be obtained by determining the values of e^x for the function $f(x) = e^x$. This is a table of values which is available in many mathematics books. If we took our $100 to a bank which offered continuous compounding of interest at a rate of 10 percent, what would be the total amount of money we would have at the end of 2 years? Sadly, it would not be equal to billions or even a paltry few million dollars. Instead we would check our table and determine the value for e raised to the exponent of 0.2 (interest rate of 0.10 multiplied by 2 years). We then retrieve the values for $e^{0.20}$. The value of this number in the table is equal to 1.2214. We thus find that the money M which is earned at the end of 2 years after a continual compounding at a 10 percent rate of an initial $100 investment is equal to $100^{(1.2214)} = \$122.14$. As we can see, continual compounding of an initial investment of $100 for 2 years will give us an entire $1.14 more than we would get by compounding that same $100 once a year for 2 years. Of course this discrepancy would add up to billions of dollars by the time the sun explodes and swallows up the earth but it will not be a monumental difference over the course of a few years. But this example does illustrate the advantage to be gained when using continuous compounding to speed up the rate at which you build your fortune.

Exponential functions also come into play when dealing with the phenomenon of population growth. Scientists and researchers who study bacteria and other microscopic beasts must often use exponential functions to calculate the rates at which these populations increase. The point of these calculations is that they can provide benchmarks to determine the effectiveness of drugs designed to eradicate these types of bacteria when introduced into their colonies. Let us say that a certain type of bacteria known as *Rosco bacillus* causes excessive, uncontrollable giggling. Under normal conditions, this bacteria will grow exponentially so that at the end of 4 days there will be 10,000 bacteria. If we test an antibacterial drug which leaves an identical colony having only 1000 bacteria at the end of 4 days, we know that the drug has an effectiveness rate of 90 percent. Clearly, we would not be able to calculate the effectiveness of the drug without knowing the amount by which it reduced the bacteria population after being introduced into the population. As a result, it would be meaningless to say that only 1000 of the *Rosco* bacteria were left after 4 days because we would have no idea as to whether we would have had 1000 or 10,000 or even more bacteria had the colony been allowed to flourish. The population count is also important because we might find that our drug actually contributed to the growth of the bacteria colony. Its tasty, lip-smacking, strawberry flavor might be just the kind of taste that *Rosco bacillus* craves for a quick energy boost or the type that would put it in an amorous mood, thereby contributing to further growth of the population. So population growth calculations provide researchers with the information they need to test both the appropriateness and effectiveness of their drugs.

How do we go about testing the rate of growth of a population of *Rosco bacillus* or any other type of population? Well, we shall save all of those time-consuming steps involved in the derivation of the exponential function for describing the rate of growth of populations and point out that in a population P of bacteria there are 1000 bacteria when our time $t = 0$. Using our skills with algebraic notation, we can express the following equation $P = 2tP_0$ where t is the number of days which have elapsed, P is the number of bacteria at time t, and P_0 is the number of bacteria at time 0. If our original bacterial population is equal to 1000, then we can substitute that value in our formula and obtain $P = 2t(1000)$. This formula tells us that our particular bacteria will increase to 2000 bacteria at the end of 1 day, 4000 bacteria at the end of 2 days, and 6000 bacteria at the end of 3 days. If we are testing the new drug Swillex against this bacteria and we find that a treated bacterial colony has 20,000 bacteria at the end of 3 days instead of 6000 bacteria, then we would conclude that we would not want to use Swillex on this type of bacteria unless we want to begin a commercial bacterial farming operation. If, on the other hand, there were only seven bacteria left at the end of 3 days after a dose of Swillex (and these few remaining bacteria were crippled, blinded, and looked far older than their age), then we could conclude that Swillex is just the thing to eradicate this particular type of bacteria.

Conclusion

Exponential functions represent merely one type of function and, indeed, both differential and integral calculus, in their most basic forms, are based upon the concept of a function—that is, the amount by which one variable changes in response to changes in another variable. The calculus is, in its most basic form, that branch of mathematics which deals with the dynamics of change. When compared to the static, almost stodgy, definition-based fields of algebra and geometry, for example, the calculus provides a framework which embraces change and motion and growth. From its very beginning, the calculus has been used to determine such things as the orbits of the planets around the sun and the rates of acceleration of freely falling bodies. More recently, it has aided efforts to understand the motions of ocean and atmospheric currents and even to construct models of entire economic systems. It is regarded as the crown jewel of mathematics and a mastery of the calculus is essential for anyone who wishes to pursue studies in advanced mathematics because it is, in the words of mathematician John von Neumann, "the first achievement of modern mathematics" which, in his opinion, "still constitutes the greatest technical advance in exact thinking."* Neumann's praise is not overstated because the calculus, as we have seen, consists of mathematical equations which can be used to solve two types of problems—those relating to the rate at which a variable quantity changes with respect to time (differential calculus) and problems where we are given the velocity of a moving body at a given instant of time and required to find the distance traveled by that body (integral calculus). Both branches of calculus have proven to be of incalculable value to scientists and engineers because they would otherwise have had no adequate tools for expressing physical laws, which entail such dynamic qualities as motion, velocity, and acceleration. Without this tool, much of what we routinely take for granted in our modern industrial society would not have been possible. Even

*John von Neumann, "The Mathematician," in James R. Newman's *The World of Mathematics*. New York: Simon & Schuster, 1956, pp. 2053–63.

though Isaac Newton, who along with the German mathematician Gottfried Wilhem von Leibniz was jointly credited with the discovery of the calculus, was not one to question his own primacy in the constellation of European science, we would be hard pressed to argue that his creation of the calculus was not as important as his discovery of his three laws of motion and his law of universal gravitation. Newton could be a bit of a snob when it came to claiming credit for his discoveries and an outright miser when forced to acknowledge the contributions of others. This arrogance did not endear him to all of his colleagues but even his most vocal opponents would probably agree that the discovery of the calculus ranks in importance with the entire edifice of Newtonian physics and has been even more impermeable than Newton's mechanistic world view which was to endure until the beginning of the 20th century.

This book has admittedly moved in fits and starts from discussions of the ways in which we gather knowledge and form our concepts of the world, through the major branches of mathematics, and finally culminating with the differential and integral calculus. We have not tried to write a textbook because there are many textbooks on calculus as well as all of the elementary branches of mathematics— algebra, geometry, and trigonometry. We have tried to present the most basic ideas without fanfare and, indeed, with a little humor here and there to show that it is possible to have fun while learning about these important mathematical ideas. No doubt most persons view mathematics with trepidation and even fear but no one has ever died from being bitten by a radical or bled to death after slicing an artery on a square root. Because mathematics is an intellectual enterprise with real-world applications, our minds are the ultimate arenas in which its principles are formulated, its equations solved, and its relevance proven.

Because this book wanders through the mathematical landscape much like a child dashing here and there on a playground, it does not lend itself to a comprehensive or systematic treatment of the subject. But that is not the intent as we would have been forced to undertake an encyclopedia-length work which would probably not appeal greatly to the lay audience. Our primary desire was to highlight the most fundamental concepts of each of the topics discussed so that we would at least provide an overview of the mathematical landscape. Although some purists may be upset that we would use such a "grubby" approach to the perfect, pristine world of algebra, geometry, and calculus, they would probably be forced to admit that this very same approach does help to illustrate the relevance of mathematical ideas to our daily affairs. Indeed, mathematics would be of very little use, no matter how perfectly logical its principles and how unblemished its proofs, if we could not relate it to the imperfect, dirty, noisy circumstances of the real world.

Bibliography

Ball, W.W. Rouse, *A Short Account of the History of Mathematics*. New York: Dover, 1960.

Bell, Eric Temple, *Men of Mathematics*. New York: Simon & Schuster, 1937.

Boyer, Carl B., *A History of Mathematics*. Princeton: Princeton University Press, 1985.

Kasner, Edward, and James R. Newman, *Mathematics and the Imagination*. New York: Simon & Schuster, 1940.

Kline, Morris, *Mathematics and the Physical World*. New York: Dover, 1959.

Motz, Lloyd, and Jefferson H. Weaver, *Conquering Mathematics*. New York: Plenum, 1991.

Motz, Lloyd, and Jefferson H. Weaver, *The Story of Mathematics*. New York: Avon, 1993.

Newman, James R., *The World of Mathematics*. New York: Simon & Schuster, 1956.

Paulos, John, *Innumeracy: Mathematical Illiteracy and its Consequences*. New York: Hill and Wang, 1988.

Smith, D. E., *History of Mathematics*. New York: Dover, 1951.

Struik, Dirk J., *A Concise History of Mathematics*, 4th ed. New York: Dover, 1987.

Index